# *From*
# SPACE TO EARTH

# *From* SPACE TO EARTH

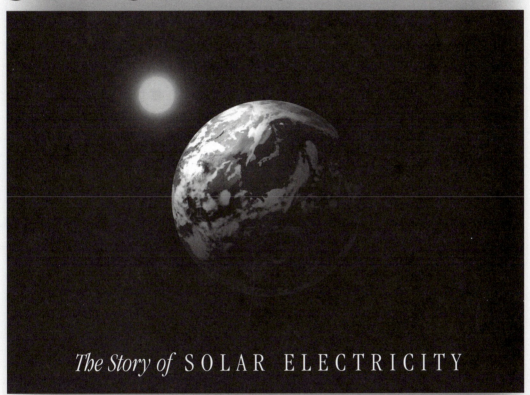

*The Story of* SOLAR ELECTRICITY

# JOHN PERLIN

*Author of* A Forest Journey
*Co-author of* A Golden Thread

aatec publications • Ann Arbor, Michigan

© 1999 by John Perlin

Published by **aatec publications**, PO Box 7119, Ann Arbor, Michigan 48107
phone & fax: 800.995.1470 (outside the United States: 734.995.1470)
email: aatecpub@mindspring.com

07  06  05  04  03  02  01  00  99          5  4  3  2  1

*Library of Congress Cataloging-in-Publication Data*

Perlin, John.
    From Space to Earth: the story of solar electricity / John
Perlin.
       p.      cm,
    Includes bibliographical references and index.
    ISBN 0-937948-14-4 (alk. paper). -- ISBN 0-937948-15-2 (pbk. :
alk. paper)
    1.  Solar power plants--History.   2. Photovoltaic power systems-
-History.   I.   Title
TH1085.P47    1999
621.31'244--dc21                                             99-23830
                                          CIP

Manufactured in the United States of America
Printed with soy ink on acid-free paper
Printed and bound by Edwards Brothers, Ann Arbor, Michigan
Jacket printed by Spectrum Printers, Inc., Tecumseh, Michigan

Cover by Hesseltine & DeMason Design, Ann Arbor, Michigan

# Contents

*During the nineteenth and early twentieth centuries, many researchers attempted to convert solar heat into power. Although technically successful, none of their approaches proved commercially sustainable. The hope of economically producing electricity with solar energy would have ended if not for the advent of photovoltaics, the direct conversion of the sun's energy into electricity by solar cells. Science magazine described the technology as "the simplest, yet most environmentally benign source of electricity yet conceived."*

*In 1876, two British scientists discovered that a certain solid material, selenium, could directly convert sunlight into electricity. One highly respected contemporary called the discovery "scientifically of the most far-reaching importance." But the inability of selenium solar cells to produce a significant amount of electricity dashed hopes that they would one day power the world's homes and industries. Nonetheless, this pioneering work proved that a solid material could change sunshine into electricity without heating a fluid and without moving parts.*

**Chapter Seven    The First Mass Earth Market**                    *57*

*The first major purchaser of solar cells for terrestrial use was the oil industry, which had both the need and the money. One need came when the government required warning lights on oil platforms in the Gulf of Mexico; most had no power other than toxic, cumbersome, short-lived batteries. Another need arose in the gas and oil fields, where small amounts of electricity are used to combat corrosion in wellheads and piping. In both instances, electrical needs existed where power lines did not. And in both situations, photovoltaics has proven indispensable.*

**Chapter Eight    Captain Lomer's Saga**                           *71*

*The U.S. Coast Guard had the ideal use for photovoltaics—powering its many isolated buoys and lighthouses. However, being a government agency, economics did not compel its leadership. A crusading Coast Guard officer, Lieutenant Commander Lloyd Lomer, believed that solar cells could save taxpayers millions of dollars. After years of objections from his commanders, Lomer finally went over their heads and received approval to develop a photovoltaics program. President Ronald Reagan commended him for "exemplary service." Today the majority of navigation aids throughout the world are powered by solar cells.*

**Chapter Nine    Working on the Railroad**                         *77*

*Microwave systems provided railways with station-to-station communication by the mid-1970s, eliminating the need for telephone lines and poles along the tracks. However, the signaling and shunting devices necessary for track safety still needed a few watts of power. Transporting even a small amount of electricity from the closest utility hook-up would cost tens of thousands of dollars, so many railroad companies turned to photovoltaics.*

**Chapter Ten    Long Distance for Everyone**                       *85*

*In the 1970s, most people in the developed world took long-distance telephone service for granted. However, residents of small communities in the western United States, isolated by rugged terrain, still had to drive long distances to make a long-distance call. Microwave repeaters could end this hardship, but they were usually located on difficult-to-access mountaintops and required some type of power. Transporting fuel or batteries to such sites was expensive and labor-intensive. In 1974, John Oades, an engineer at a GTE subsidiary, solved the problem by designing an extremely low-powered repeater that could run solely on photovoltaics. Australia, a large country with a relatively small population, began building photovoltaic-powered communication networks as early as 1978 to service its widely dispersed citizenry. By the mid-1980s, solar cells had become the energy source of choice for remote telecommunications networks worldwide.*

*Photovoltaic-generated electricity now costs around 25¢ per kilowatt hour. This is less expensive than any other stand-alone power source, such as batteries or diesel generators, and it is less expensive than installing transmission lines underground anywhere in the world or stringing overhead lines over 250 yards to a small power user. At this writing, however, solar electricity still costs more than power from existing overhead utility lines. Many believe that only by devising new ways to make solar cells will their price drop low enough to compete with utility power. Currently, most cells are made from relatively thick crystalline silicon. Thin-film photovoltaic materials now appearing on the market—amorphous silicon, cadmium telluride, copper indium diselenide, thin-crystalline silicon, and sheet-crystalline silicon—may provide the solution.*

*The reliability and versatility of photovoltaics from space to earth has impressed many in the power and telecommunications industries. Twenty years ago, institutions such as the World Bank knew little about photovoltaics. Now the Bank and many other world power brokers view solar cells as having "an important and growing part in providing electrical services in the rural areas of the developing world." In telecommunications, these same agencies believe that only photovoltaics offers "a real practical possibility of reliable rural telecommunication for general use." Opportunities in the developed world abound as well. The growing concern over global warming promises to transform photovoltaics into a major energy producer, allowing everyone the benefits of electricity without doing harm to our home, planet earth.*

# Preface

Four years ago, I was asked to write an updated version of *A Golden Thread: 2500 Years of Solar Architecture and Technology*, which I co-authored with Ken Butti. *A Golden Thread* examined the major advances in solar energy, from the time of the ancient Greeks to the late 1970s. Nearly twenty years later, as I began to look at the direction solar technologies were headed, every aspect seemed in decline. Gone were the heady days when worshipping crowds chanted "Solar Yes! Nuclear No!" The sales of solar water heaters had plummeted. The firm that built the Luz solar power generating plant, a massive field of mirrors in the Mojave desert, once viewed as the precursor to vast solar electric stations throughout the world, had gone bankrupt. And energy prices, expected to rise to astronomical levels, were lower than ever.

I began to wonder if there was anything significant to write about. But when I began to investigate photovoltaics, the direct conversion of sunlight—not sun heat—into electricity by solar cells, I found that here was a solar success story that begged to be told. Were it not for photovoltaics, those long-lived satellites so necessary for science, the military, and the information superhighway would not have been possible. Photovoltaics has brought abundant clean water, electric lighting, and telephone service to those who had gone without. It ensures the safe passage of ships and trains, powering navigation aids and warning devices. In fact, the use of solar cells already benefits hundreds of millions of people throughout the world, though few have any knowledge of this.

In the twenty-five years since the market for earth-based applications of solar cells opened, production, revenues, and uses have grown phenomenally. In the spring of 1973, the annual output of photovoltaics was around twenty to thirty kilowatts; the next year it doubled.[1] In 1978, the industry reached the one-megawatt mark.[2] By the mid-1990s, the small country of Mongolia alone had two megawatts powering its electrical systems.[3] World production exceeds one hundred megawatts.

The upward trend in production has brought great increases in revenue. In 1978, the fledgling solar cell industry grossed a mere 12.5 million dollars.[4] Nineteen years later, that figure was close to two billion dollars, with industry growth at a healthy rate of 24 percent per year.[5]

Although only fifteen megawatts of solar cells were produced in Europe in 1996, the projection is to produce seven times that amount before 2000. Production worldwide could reach three hundred megawatts by the end of the millennium.[6] This phenomenal success has led many to envision that "the direct harnessing of solar energy by photovoltaic cells has the potential to become a major $CO_2$-free energy source."[7]

*From Space to Earth: The Story of Solar Electricity* covers the milestones of photovoltaics and its uses—from the technology's shaky beginnings in the late nineteenth century, mired in scientific controversy, to its current position as an enduring power source that is improving the daily lives of many. However, even as solar cells seep into everyday life, few realize their presence. The step-by-step progress of photovoltaics has elicited little fanfare. It is my hope that *From Space to Earth* will end the silence and give this amazing technology its rightful place in the sun.

## Notes

1. H. Kelly, "Photovoltaic Power Systems, *Science* 19 (10 February 1978): 635.
2. T. Kidder, "Solar Power—Somewhere over the Rainbow," *Atlantic Monthly* 245 (June 1980): 69.
3. D. Agchbayar, personal correspondence, November 10, 1996.
4. Kidder, "Solar Power."
5. AstroPower Prospectus, February 12, 1998, 3.
6. *Solar Europe*, December 1, 1997. (Brochure provided by Dr. Wolfgang Palz.)
7. B. Andersson et al., "Material Constraints for Thin-Film Solar Cells," *Energy* 23, no. 5 (1998): 407.

# Acknowledgments

To accurately and thoroughly tell a story never before told required researching through much archival material and interviewing many people key to photovoltaics and its uses. I would like to thank the following individuals and institutions for their generous assistance.

*Chapter 1:* The Butti–Perlin Archives, Santa Barbara, California. *Chapter 2:* Mary Sampson, Archivist, Royal Society, London; Vicki Holtby, Archives Assistant, Kings College London; Adam J. Perkins, Royal Greenwich Observatory Archivist, Department of Manuscripts and University Archives, Cambridge University Library; Kate Manners, Assistant Archivist, Manuscripts and Rare Books, University College London. *Chapter 3:* Dr. Sheldon Hochheiser, Director, AT&T Archives, Warren, New Jersey, and his staff; Audrey Chapin Svensson; the late Joseph Loferski. *Chapter 4:* Dr. Martin Wolf, Professor Emeritus, Electrical Engineering, University of Pennsylvania. *Chapter 5:* Command Historian Richard Bingham, U.S. Signal Corps, Ft. Monmouth, New Jersey; Dr. Hans Ziegler. *Chapter 6:* Dr. Elliot Berman; Ed Mahoney; Bernard McNelis; Arthur Rudin; Clive Capps. *Chapter 7:* Guy Priestley; Harry Sanger; Steve Trenchard; Ed Mahoney; Larry Beil. *Chapter 8:* John Goldsmith; Captain Lloyd Lomer (USCG Ret.). *Chapter 9:* Arthur Rudin; Robert Mitchell; James Le Vere; Stanley Taylor. *Chapter 10:* John Oades; William Hampton; Michael Mack; Arnold Holderness. *Chapter 11:* Stephen Allison; the late Gottfried Kleiber; Jerome Billerey; Guy Oliver; Dr. Wolfgang Palz; Dominique Campana; Bill Yerkes. *Chapter 12:* Jim Martz; Patrick Jourde; Richard Acker; Mark Hankins; Richard Hansen; Neville Williams. *Chapter 13:*

Steve Strong, for graciously introducing me to European leaders in on-site residential and commercial photovoltaics; Dr. Markus Real; Joachim Benemann; Donald Osborn; Gregory Kiss. *Chapter 14:* Fritz Wald; Dr. David Carlson; Dr. Christopher Wronski; Dr. Allen Barnett; Dr. Martin Green; Dr. Stuart Wenham; Dr. Richard Bleiden; Dr. Subhendru Guha. *Chapter 15:* Stephen Allison; Terry Hart; Bill Yerkes; Roland Skinner; Dr. Markus Real; Dr. Charles Keeling; Dr. Tom Wigley; William Young; James Trotter.

In addition, I would like to thank Darla Anderson, Duane Anderson, Paul Basore, Trish Carrico, Paul Caruso, Momadov Diarra, Mark Farber, Mark Fitzgerald, Charles Gay, Richard Gauld, William Gould, Lalith Gunaratne, Gary Johnson, Dr. Daniel Kammen, Carl Kotiela, Roxanne Lapidus, Peter Lawley, Hugues Le Masson, Dr. Mel Manalis, Mark O'Neill, John Quinney, Vella Rivera, Lou Shrier, Ken Stokes, Richard Swanson, Steve Taylor, John Thornton, Mark Trexler, Rob Van der Plas, Larry Weingarten, Bob Willis, and the University of California at Santa Barbara Davidson Library.

I give thanks to three of the world's most respected authorities on photovoltaics for their review of the manuscript. Peter Iles, who has spent forty years in the research and development of solar cells for space use, and who is recognized by his peers for his vast knowledge of the history of photovoltaics; Bernard McNelis, who began his engineering career in 1973, just as terrestrial applications of solar cells began, and is now one of the most highly regarded photovoltaics consultants; and Dr. Christopher Wronski, Leonhard Professor of Micro Electronic Devices and Materials, Pennsylvania State University, who spent decades in solar cell research and development at Exxon and RCA Laboratories, and is the co-discoverer of the Wronski/Staebler effect.

I am also fortunate to have as a friend another authority on photovoltaics, Dr. Richard Komp, who has been involved in photovoltaics for over thirty years, starting with the late Dr. Dan Trivich at Wayne State University. I thank him for his many valuable recommendations.

I thank the following nontechnical reviewers for their meticulous work on the manuscript, making it more accessible to the general reader: Jan Corazza; Luciano Corazza; Barbara Francis; Christina Bych; and, my wife, Margaret November, MD.

Finally, I would like to thank the Department of Energy for its generous financial assistance.

Chapter One

# Photovoltaics:
# The Great Solar Hope

Alfred Russel Wallace, along with Charles Darwin, gained great fame and respect as a co-discoverer of the theory of evolution. In his book *The Wonderful Century*, Wallace described the important technological developments that occurred during his lifetime, which spanned almost the entire nineteenth century. He believed that these changes significantly separated his period from the rest of world history. For example, Wallace noted that from the earliest times until the mid-nineteenth century, land transportation had not changed at all. People either walked or relied on animals to carry them and their goods overland. "The speed for long distances must have been limited to ten or twelve miles [sixteen to nineteen kilometers] an hour" at best, Wallace contended. Whether "ancient Greek or Roman, Egyptian, or Assyrian, [or early nineteenth-century] Englishman," all traveled about "as quickly and as conveniently." Then what Wallace termed an "entirely new departure" in transportation occurred. "Railroads raised the speed of transport to fifty or sixty miles [eighty to ninety-five kilometers] per hour," revolutionizing travel and the conveyance of goods."[1]

"In . . . navigation," Wallace saw, "a very similar course of events." For thousands of years, people had to depend on oars and sails. He judged even "the grandest three-decker or full-rigged clipper ship but a direct growth . . . from the rudest sailing boat of the primeval savage. Then, at the

*Alfred Russel Wallace*

*The frontispiece of Augustine Mouchot's ground-breaking book that described solar applications past and present to his nineteenth-century audience.*

very commencement of the present century, the totally new principle of steam propulsion began to be used," hailed by nineteenth-century observers as "without parallel in the history of man as regards to commerce and rapidity of communications."[2] Indeed, prior to the steamboat, the best way to navigate from Pittsburgh to New Orleans was by keelboat. A round trip took six to seven months. The steamer reduced the voyage to a little over three weeks.[3]

Likewise, engines propelled by steam unleashed industry. No longer tied to water power, factories proliferated, producing for mass consumption a plethora of goods hitherto confined to the wealthy.

These great steps, heralded as "man's increased power over nature," came with a steep price.[4] Most of the engines that propelled these locomotives, ships, and machines burned coal—and the amount of coal they consumed was alarming. In the late 1800s, one leading French engineer worried, "We are currently spending the supplies of energy accumulated over the millions of centuries. Industry is devouring this savings account . . . and one wonders how much longer we can borrow against it."[5] Another nineteenth-century Frenchman, Augustine Mouchot, a professor of mathematics at the Lycée de Tours, expressed even greater anxiety over the situation, prophesying, "Eventually industry will no longer find in Europe the resources to satisfy its prodigious expansion. . . . Coal will undoubtedly be used up. What will industry do then?"[6]

Mouchot thought that perhaps the sun's heat could replace the burning of coal to run Europe's industries. He therefore first studied what had already been done to put solar energy to use. When he learned that the excavators of Pompeii had unearthed window glass similar to that used in his own time, he speculated that the Romans had discovered that clear glass exposed to the sun acts as a solar heat trap. Ancient texts have proven Mouchot correct. The Romans found from experience that when sunlight enters a structure and strikes the floor and walls, it transforms into heat, which cannot easily exit through glass. They named these sunspaces *heliocamini* or "sun furnaces."[7] Scientists today use the term "greenhouse effect."

Like so much the ancients had developed, window glass did not survive the fall of Rome. It was not until the Renaissance that glass became as common as it had been in ancient Rome. People then again realized, as Horace de Saussure, one of eighteenth-century Europe's foremost naturalists, observed, "that a room, a carriage, or any other place is hotter when the rays of the sun pass through glass."[8]

*A large Roman window facing south to collect solar heat.*

Because no one really knew just how much solar heat glass could trap, Saussure took it upon himself to find out. In 1767, he built a miniature greenhouse by stacking five glass boxes of increasing size one inside the other on a black wooden table. Despite the mild weather the day of his experiment, the bottom of the innermost box heated to 190°F (88°C). By replacing the glass sides with wood insulated by black cork, the bottom box heated to 228°F (109°C)—16°F above the boiling point of water.

*A simplified version of Saussure's glass-covered boxes. In the early 1880s, Samuel Pierpont Langley, secretary of the Smithsonian Institution, carried the device to the top of Mt. Whitney to study the sun and its effects in high altitudes.*

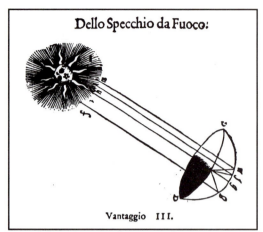

*A Renaissance drawing of a concave mirror focusing the sun's heat.*

The simplicity of Saussure's device, combined with the fact that it could collect more than enough heat to boil water, fit with Mouchot's ambition to drive industrial steam engines by the sun. However, after much calculation and analysis, Mouchot concluded that to produce enough electricity to actually power machinery, his solar plant would have to be so large that it would not be feasible, either in cost or practice.

But he did not give up. Further studies showed that concave reflectors could optimize solar heat collection. The ancients had called such reflectors "burning mirrors" because they focused the sun's rays to an intensity that could burn wood and melt metals. Through reading old texts, Mouchot learned that two hundred years earlier a fellow Frenchman named Villette had made a burning mirror that, as one eyewitness reported, created a flame "most forcibly of any fire we know."[9] The observer, an English traveler, thought that if these powerful mirrors were redesigned, they "would be of great use," especially to England's iron industry, whose potential was

*One of Augustin Mouchot's solar reflectors drove a steam engine at the Universal Exposition in Paris in 1878.*

held back for lack of wood fuel. Mouchot's work with mirror technology led him to develop the first sun motor, which produced sufficient steam to drive machinery. One journalist described it as a "mammoth lamp-shade, with its concavity directed skyward."[10]

Mouchot's success aroused much debate in France during the 1870s and 1880s. Many were convinced that solar energy could produce unlimited power at almost no cost. Others dismissed his device as nothing more than a toy. To resolve the controversy, the government set up a commission to investigate the matter. After a year of testing Mouchot's sun machine, the commission ruled, "In France, as well as in other temperate regions, the amount of solar radiation is too weak for us to hope to apply it . . . for industrial purposes."[11]

However, the report did leave hope for future solar inventors by adding, "Though the results obtained in our temperate and variable climate are not very encouraging, they could be much more so in dry and hot regions where the difficulty of obtaining other fuel adds to the value of solar technologies. . . . [Hence,] in certain special cases, these solar apparatuses could be called upon to provide useful work."[12]

The nineteenth-century Swedish–American engineer John Ericsson, who probably contributed more to the ascendancy of steam power than any other individual on earth, also argued that the sunny areas of the world were the place for sun-driven machinery. In his words, "The application of the solar engine [in these regions] is almost beyond computation while the source of its power is boundless."[13]

Although Ericsson's screw propeller had assured the supremacy of steamship over sail, and the locomotive he built, one of the first, helped to usher in the age of railroading, the steam engine's huge appetite for coal came to haunt him. He feared, as had Mouchot, that "the time will come when . . . Europe must stop her mills for want of coal."[14] Furthermore, Ericsson shared Mouchot's belief that solar power offered the only way to avert an eventual global economic paralysis that would result in putting "a stop to human progress."[15] The inventor felt such an urgency to develop solar engines that he devoted the last two decades of his life to this pursuit.

*John Ericsson, the inventor of the* Monitor, *the first iron-clad vessel.*

*One of John Ericsson's sun machines.*

Ericsson was elated when he completed his first working model, telling a friend, "It marks an era in the world's mechanical history."[16] But three years and five experimental engines later, Ericsson's enthusiasm had been tempered. His experiments taught him that "although the heat is obtained for nothing, so extensive, costly, and complex is the concentrating apparatus" that engines powered by solar energy were actually more expensive than coal-fueled motors.[17]

The failure of Mouchot and Ericsson to come up with economical solar machinery did not dampen everyone's enthusiasm for solar power. In fact, one energy specialist wrote in 1901, "The solar engine is [still] exciting special interest."[18] The truth of that statement could be seen in Frank Shuman's career. Described in a 1909 issue of *Engineering News* as a "man of large practical experience,"[19] Shuman felt, as had Mouchot and Ericsson, the dire need for solar-run machinery if the world were to continue its industrial development. But he did not wish to repeat their mistakes.

After studying the sun machines invented by his predecessors, he found that cost was "the rock on which, thus far, all sun-power propositions were wrecked."[20] He therefore turned his back on reflectors and focused his attention on glass, just as Mouchot had originally done.

In his backyard in the Philadelphia suburb of Tacony, Shuman laid over one thousand square feet (ninety-three square meters) of glass-covered blackened pipes in which a liquid with a low-boiling point circulated. The solar-heated vapor operated an engine, which demonstrated "[t]he practical possibility of getting power from sun heat by the 'hot bed' plan."[21]

The "thousands of barrels of water" that his plant pumped under Pennsylvania's summer sun gave Shuman confidence that one day his sun machines would make agriculture and industry possible in the fuel-shy but sunny regions of the world.[22] However, first he had to cross the same Rubicon that had held Mouchot back: A glass-covered solar plant needed a much greater surface area than the comparatively compact coal-fired engine to produce the same amount of power.

But Shuman found a way to cleverly avoid past pitfalls. First, he chose to locate his commercial solar motor in Egypt, where land and labor were cheap, sun plentiful, and coal costly. Second, to increase the amount of heat generated by the solar motor, the glass-covered pipes were cradled at the focus of a low-lying troughlike reflector. A field of five rows of collectors was laid in the Egyptian desert. The solar plant also boasted a storage system, which collected excess sun-warmed water in a large insulated tank for use at night and during inclement weather. In contrast, solar systems that lacked storage, like those designed by Mouchot and Ericsson, that "go into operation only when the sun comes out from behind a cloud and go out of action the instant it disappears again can hardly be expected to pay dividends," an engineer familiar with early solar machinery stated.[23] For industrialists, Shuman's solution eliminated a major obstacle to solar's appeal.

*Frank Shuman, solar inventor.*

Shuman's solar plant far surpassed the performance of all previous solar engines and, in Egypt, proved more economical than a coal-fired plant. Engineers recognized that Shuman's breakthrough demonstrated that "solar power was quite within the range of practical matters."[24] Even former skeptics, like those at *Scientific American*, now praised Shuman's solar engine as "thoroughly practical in every way."[25]

But all the hopes and plans of solar engineering disintegrated with the outbreak of World War I. The staff of the Egyptian plant had to leave for war-related work in their respective homelands. Shuman, the driving force behind large-scale solar development, died before the war ended. Worse yet, after the war the world turned to oil to replace coal. Oil and gas reserves were found in sunny, coal-shy regions like southern California, Iraq, Venezuela, and Iran—places that had been targeted by Shuman, as well as Mouchot and Ericsson, as prime locations for solar plants. With oil and gas selling at near-giveaway prices, scientists, government officials, and businessmen became complacent over the world's energy situation. Interest in sun power came to an abrupt end.

In the fifty years that followed, oil's low price and seemingly endless supply kept any serious commercial solar activity at bay. Then came the

*Shuman's first solar motor, which used glass to trap solar heat, ran this pump in suburban Philadelphia.*

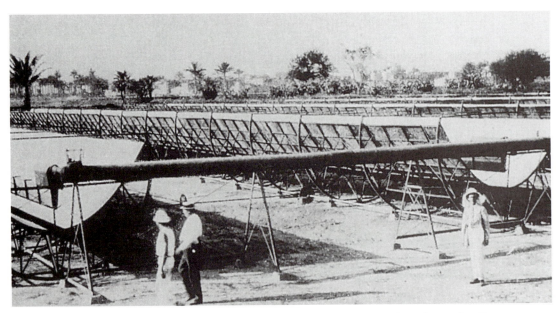

*Rows of trough reflectors powered Shuman's successful sun machine located in Egypt.*

great wake-up call, the oil shock of late 1973. As a result of the embargo imposed by the Arab oil-producing states, the price of oil quadrupled and supplies dwindled. For the next seven years, it continued to climb. In 1980, a barrel of oil cost $32, a 25-fold increase from 1970.[26] Once again, more costly and increasingly scarce conventional fuel supplies made the world look to solar energy. Since it was feared that the utilities would "be severely affected by any ensuing shortfall in gas and oil supplies," work began in earnest to harness the sun for electrical power.[27] The power tower was one option considered.

The power tower concept resembles the use of solar heat that ancient writers ascribed to the mathematician Archimedes. To defend his native Syracuse from the ravages of the invading Roman fleet, Archimedes arranged a number of flat mirrors so that they concentrated the incoming sunlight upon the wooden hulls of the enemy ships "to kindle a fearsome fiery heat . . . [which] reduced them to ashes."[28] Power towers also rely on the proper placement of a number of flat mirrors. For example, a one hundred-megawatt plant would require twenty-five thousand mirrors on four hundred acres of land! But instead of trying to burn ships, these mirrors would move throughout the day to focus sunlight onto a boiler atop a 735-foot tower.[29] The heat produced by the concentrated sunlight would reach around 1000°F (538°C) and so run a conventional turbogenerator.

However, power towers have yet to emerge from the government-funded demonstration-project stage to the commercial market, despite the hundreds of millions of dollars spent on their research and development. And, too, the introduction of a simpler, less futuristic alternative took the wind out of the power towers' sails. This technology uses trough-shaped reflectors that look hauntingly similar to the parabolic troughs Frank Shuman built in Egypt seventy years earlier. Its story has eerily paralleled Shuman's Egyptian experience, too. Experts at first called the Luz trough-reflector plant, built in the California desert, "cost competitive with conventional fuel power stations" and declared its "potential . . . virtually unlimited."[30] Savvy investors confidently signed a contract which guaranteed that the local utility would buy solar-generated electricity at the same price it would have had to pay for fossil fuels. With oil close to $40 a barrel in the early 1980s, and additional escalations in price predicted, it is no wonder that they had expected to make a lot of money. But just like Shuman, the investors were in for a surprise. The price of fossil fuels hit rock bottom in the late 1980s and,

*A field of flat mirrors focuses sunlight onto a tower to produce steam for a conventional turbogenerator.*

*Although the company that built this modern trough-reflector plant in the Mojave Desert went bankrupt, the plant still produces megawatts of power.*

again, it appeared there was no end to the supply. As a consequence, the company responsible for building the plant went bankrupt.[31]

"This looks like the end of the story," to quote Beatrix Potter at the moment Old Brown took out his knife to skin Squirrel Nutkin. However, just as Potter let her readers know, ". . . it isn't."[32] Scientists have devised a radically different technology for exploiting solar energy for power, one that casts aside the notion of using the sun's heat. The technology is photovoltaics, the direct conversion of the sun's energy into electricity via fragile-looking solar cells, no more than several hundred microns thick. Hardly the rugged stuff utility people are accustomed to, photovoltaics does away with the bulky paraphernalia—boilers, turbines, pipes, and cooling towers—required by all other electricity-generating technologies. In fact, solar cells operate without moving parts. Within those few microns, photons, packets of energy from the sun, silently push electrons out of the cells and so make electricity. As Markus Real, who in the late 1970s began his engineering career searching for ways to transform solar heat to electricity, recalled, "For me it was clear when I saw the two technologies [solar reflectors and photovoltaics]. I realized that solar cells were something new, something special. I was convinced we had here a revolution in power generation."[33]

Real was not the only person excited by photovoltaics. Around the same time, staff writers for *Science* magazine declared, "If there is a dream solar technology, it is photovoltaics—solar cells . . . a space-age electronic marvel at once the most sophisticated solar technology and the simplest, most environmentally benign source of electricity yet conceived."[34]

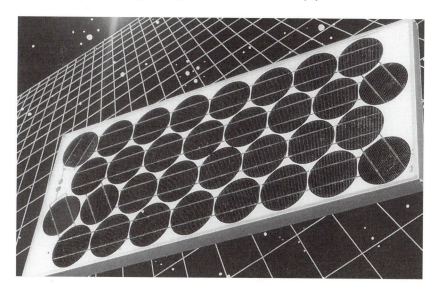

*These disks are solar cells. Interconnected and framed, they form a photovoltaic panel or module. Despite their ethereal appearance, when exposed to the sun, solar cells produce electricity that is no different than that generated by huge turbines.*

## Notes & Comments

1. A. Wallace, *The Wonderful Century: Its Successes & Failures* (London: Swan Sonnenschain, 1898), 2, 3, 6–7. Dr. Jay Stephen Gould's column, "The View of Life," in the September 1998 issue of *Natural History* led me to look at Wallace's book.

2. Ibid., 8;  H. Tudor, *Narrative of a Tour in North America* (London: J. Duncan, 1834), 2.36.

3. J. Perlin, *A Forest Journey* (Cambridge, MA: Harvard University Press, 1991), 342.

4. Wallace, *The Wonderful Century*, 2.

5. M. Crova, "Rapport sur les Experiences Faites a Montepellier pendant l'Anne 1881 par la Commission des Apparelis Solaires," *Académie des Sciences et Lettres de Montepellier, Section des Sciences* 10 (1884): 289–90.

6. A. Mouchot, *La Chaleur Solarie et ses applications industrielles*, 2nd ed. (Paris: Gauthier-Villars, 1879), 256.

7. E. A. Andrews, *A Latin Dictionary* (Oxford: Clarendon Press, 1879), 845.

8. H. de Saussure, "Lettre de M. de S aux Auteurs du Journal," *Le Journal de Paris*, Supplement, au #108, (17 April 1784), 475. *Also see* H. de Saussure in M. Achille Comte de Buffon, ed., *Oeuvres Completes de Buffon, Tome Premiere* (Paris: Bazouge-Pigoreau, 1839), 183n.

9. "An Account from Paris Concerning a Great Metallin [sic] Burning Concave. . ." in *Philosophical Transactions* V, no. 47 (19 July 1669): 986–87.

10. L. Simonin, "L'Emploi Industriel de la Chaleu Solaire," *Revue des Deux Mondes* (1 May 1875): 204.

11. Crova, "Rapport sur les Experiences," 323.

12. Ibid., 325–26.

13. John Ericsson, in W. Church, *The Life of John Ericsson*, vol. 2 (New York: C. Scribner and Sons, 1890), 266.

14. J. Ericsson, *Contributions to the Centennial Exposition* (New York: Printed by the author, 1876), 577.

15. Ericsson, in Church, *The Life of John Ericcson*, 191.

16. Ericsson, *Contributions to the Centennial Exposition*, 564.

17. Ericsson, in Church, *The Life of John Ericcson*, 271.

18. R. Thurston, "Utilizing the Sun's Energy," *Smithsonian Institution Annual Report* (1901) 265.

19. "Power from the Sun's Heat," *Engineering News* 61, no. 19 (13 May 1909): 509.

20. F. Shuman, *The Generation of Mechanical Power by the Absorption of the Sun's Rays* (Tacony, Philadelphia: By the author, 1911), 2.

21. "Power from the Sun's Heat," 509.

22. Ibid.

23. Thurston, "Utilizing the Sun's Energy," 268.

24. H. Jenkins, in A. Ackermann, "The Utilisation of Solar Energy," *Journal of the Royal Society of Arts* 63 (1915): 564.

25. F. Shuman, "The Feasibility of Utilizing Power from the Sun," *Scientific American* 110 (25 February 1914): 179.

26. J. Lindmayer, "Industrialization of Photovoltaics," in *Third E.C. Photovoltaic Solar Energy Conference* (Cannes, France) (Dordrecht: Kluwer Academic Publishers, 1981), 179.

27. A. Hilderbrandt and S. Dasgupta, "Survey of Power Technology," *Journal of Solar Energy Engineering* 20 (20 May 1980): 91

28. Diodorus Siculus, *Diodorus of Sicily* (Cambridge, MA: Harvard University Press, 1933–1967), XXXVI.18.1.

29. "How Solar Technologies Will Work," *Business Week* (9 October 1978): 96.

30. "The Luz Projects," *Sunworld* 9, no. 4 (1985): 111–12.

31. The Luz plant remains in operation and still generates electricity.

32. Beatrix Potter, *The Tale of Squirrel Nutkin* (New York: F. Warne & Co., 1903), 50.

33. Markus Real, 1994, videotape. (Courtesy Mark Fitzgerald.) Markus Real was not the only defector from power towers to photovoltaics. Bernard McNelis, managing director of IT Power, one of the world's foremost renewable energy consulting groups, spent several years in the late 1970s working on the first European power tower. Although he described his involvement as "very exciting from an engineering/scientific viewpoint, and dealing with European Community officials and French, German, and Italian contractors challenging," the shortcomings of the technology "convinced [him] that [he] had to get back to photovoltaics," where he had begun his solar career. Correspondence with Bernard McNelis.

34. A. Hammond, "Photovoltaics: The Semiconductor Revolution Comes to Solar," *Science* 197 (23 July 1977): 444.

<br />

## Chapter Two
# Photoelectric Dreams

The direct ancestor of today's solar cell got its start, along with the information highway, in the last half of the nineteenth century, when investors poured enormous sums of money into the construction of a seamless worldwide telecommunications network. Cables were laid undersea and wires were strung overland so that people separated by great distances and geographical barriers—mountains and oceans—could converse with one another instantaneously, first by telegraph and later by telephone.

While laying the trans-Atlantic telegraph cable in the 1860s, Willoughby Smith, the project's chief electrician, invented a superior device for detecting flaws in the cables as they were submerged. Searching for an inexpensive material to use in the testing apparatus, Smith tried bars of crystalline selenium.[1] Superintendent May, who oversaw the testing, reported that although the selenium bars worked well at night, they failed dismally when the sun came out.[2]

Suspecting that selenium's peculiar performance had something to do with the amount of light falling on it, Smith placed the bars in a box with a sliding cover. When the box was closed and light excluded, the bars' resistance—the degree to which they hindered the electrical flow through them—was at its highest and remained constant. But when the

<br />

cover of the box was removed, their conductivity—the enhancement of electrical flow—immediately "increased from 15 to 100 percent, according to the intensity of light."[3]

To determine whether it was the sun's heat or its light that affected the selenium, Smith conducted a series of experiments. In one, he placed a bar in a shallow trough of water. The water blocked the sun's heat, but not its light, from reaching the selenium. When he covered and uncovered the trough, the results obtained were similar to those previously observed, leading him to conclude, "The resistance [of the selenium bars] was altered . . . according to the intensity of light to which they were subjected."[4]

As a consequence of Smith's publishing an account of his trials and tribulations with selenium, one of Europe's leading scientists remarked in 1876, "The attention of physicists is at present very much directed to" the material.[5] Among the researchers examining the effect of light on selenium were the British scientist Professor William Grylls Adams and his student Richard Evans Day. During the late 1870s they subjected selenium to many experiments, including one in which they passed a battery-generated current through it.

*William Grylls Adams, who, with his student, Richard Evans Day, discovered the photoelectric effect in a solid material.*

After the selenium was detached from the battery, Adams and Day discovered to their surprise that the current running inside the selenium had reversed itself. To find out why the selenium had changed the direction of the electrical flow, they repeated the experiment with one variation. After removing the selenium from the battery, they let a flame shine onto the selenium. The flame forced the current to flow in the direction opposite to that in the previous experiment. "Here there seemed to be a case of light actually producing an electromotive force within the selenium, which in this case was opposed to and could overbalance the electromotive force" of the battery, the amazed scientists observed.[6]

This unexpected result led Adams and Day to alter their course of investigation and to immediately examine "whether it would be possible to start a current in the selenium merely by the action of light."[7] The next morning, they lit a candle an inch away from the same piece of selenium. The needle to their measuring device reacted immediately. Screening the selenium from light caused the needle to drop to zero. These rapid responses ruled out the possibility that the heat of the candle flame had produced the current (a phenomenon known as thermal electricity), because when heat was applied or withdrawn in thermoelectric experiments, the needle would always rise or fall slowly. "Hence," the investigators concluded, "it was clear that a current could be started in the selenium by the action of the light alone."[8] They therefore felt confident that they had discovered something completely new: that light caused "a flow of electricity" in a solid material. Adams and Day called current produced by light "photoelectric."[9]

A few years later, Charles Fritts of New York moved the technology forward by constructing the world's first photoelectric module. He spread a large, thin layer of selenium onto a metal plate and covered it with a thin, semitransparent gold-leaf film. The selenium module, Fritts reported, produced a current "that is continuous, constant, and of considerable force . . . not only by exposure to sunlight, but also to dim diffused daylight, and even to lamplight." As to the usefulness of his invention, Fritts optimistically predicted that "we may ere long see the photoelectric plate competing with [coal-fired electrical-generating plants]," the first of which had been built by Thomas Edison in 1882, just three years before Fritts announced his intentions.[10]

Fritts sent his solar panels to Werner von Siemens, whose reputation ranked alongside Edison's among those working with electricity and whose experiments with light and sele-

A handwritten page from the manuscript Adams and Day published about their important discovery, including a drawing by Adams of the first selenium solar cell.

*Werner von Siemens, the first major scientist to realize the significance of the photoelectric effect and to urge its study by the scientific community.*

nium had been published in the world's leading scientific journals. The panels' output of electricity when placed under light so impressed Siemens that the renowned German scientist presented Fritts' devices to the Royal Academy of Prussia. Siemens declared to the scientific world that the American's modules "presented to us, for the first time, the direct conversion of the energy of light into electrical energy."[11]

Siemens judged photoelectricity to be "scientifically of the most far-reaching importance."[12] An even greater scientist of the time, James Clerk Maxwell, agreed. He praised the study of photoelectricity as "a very valuable contribution to science." But neither Maxwell nor Siemens had a clue as to how the phenomenon worked. Maxwell wondered, "Is the radiation the immediate cause or does it act by producing some change in the chemical state of the [selenium]?"[13] Siemens did not even venture an explanation, but urged a "thorough investigation to determine upon what the electromotive light-action of [the] selenium depends."[14]

Few scientists heeded Siemens' call because most of his contemporaries viewed photoelectric devices, such as Fritts' "magic" plates, as perpetual motion machines. They appeared to generate power without consuming fuel and without dissipating heat. Any Victorian worth his salt knew that "[s]uch efforts could never succeed."[15]

Several courageous scientists, however, charged that their profession's denigration of photoelectricity was based on ignorance. For example, George M. Minchin, a professor of applied mathematics at the Royal Indian

*Fritts' selenium solar module.*

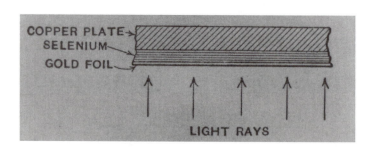

Engineering College, complained to a scientist friend that rejecting photoelectricity as scientifically unsound originated from the "very limited experience" of contemporary science and "from a 'so far as we know' [perspective, which] is nothing short of madness."[16] In fact, Minchin came closest among the handful of nineteenth-century experimentalists to explaining what happened when light strikes a selenium solar cell. Perhaps, Minchin wrote, it "simply act[s] as a transformer of the energy it receives from the sun, while its own materials, being the implements used in the process, may be almost wholly unmodified."[17]

The scientific community during Minchin's time also dismissed photoelectricity's potential as a power source by looking at the results obtained when measuring the sun's thermal energy in a glass-covered black-surfaced device, the ideal absorber of solar heat. "But clearly the assumption that all forms of energy of the solar beam are caught up by a blackened surface and transformed into heat is one which may possibly be incorrect," Minchin argued.[18] In fact, he believed, "There may be some forms of [solar] energy which take no notice of blackened surfaces [and] perhaps the proper receptive surfaces" to measure them "remain to be discovered."[19] Minchin intuited that only when science had the ability to quantify "the intensities of light as regards each of [its] individual colours [that is, the different wave lengths]" could scientists judge the potential of photoelectricity.

Albert Einstein shared Minchin's suspicions that the science of the age failed to account for all the energy streaming from the sun. In a dar-

The max. e.m.f. develop.d by candle takes about 7 or 10 minutes to show itself.
The fall is slower (except at first); but the residual effect can be pumped out of the battery by a few seconds' passage of a small current from a Daniell, & the Photo battery starts fresh.
No amt. of exposure to light enfeebles the action. The batteries seem to keep for ever.
[Obs: No current passes thro' the Photo battery. Its poles are connected with a Thomson Quad.r Elect.r]



The formation of the sensitive plates is a troublesome process, & succeeds in only about 5 or 6 out of 10 trials.
The liquid is a certain kind of alcohol.

I'm making a shot at the theory of [illegible] cells. It seems clear that a Daniell current is sent thro' a beaker with [illegible] plates in $H_2O$, then = for the inverse e.m.f. of the cell is of the form

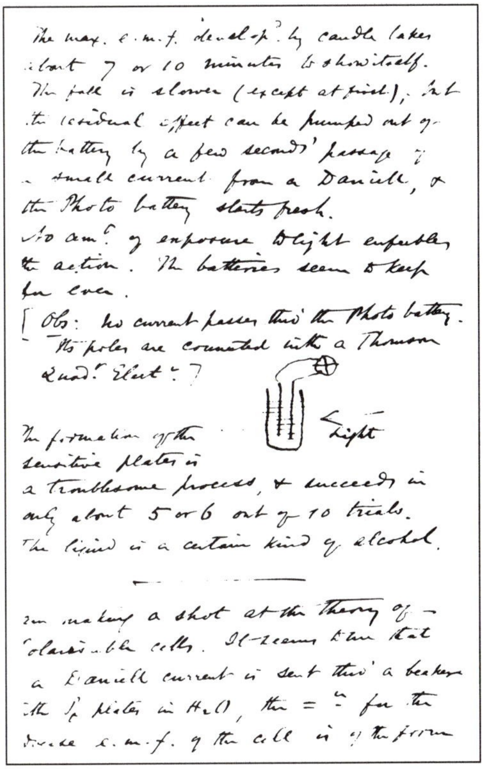

*A letter written by George Minchin in which he discusses his photoelectric work. The illustration shows his first experiment in solar electricity. Minchin placed two separate tin plates in a glass tube filled with alcohol, exposed it to light, and thereby produced electricity. He had planned to patent the discovery, until he learned that Edmond Becquerel had conducted similar experiments forty-one years before.*

ing paper published in 1905, he showed that light possesses an attribute that earlier scientists had not recognized. Light, Einstein discovered, contains packets of energy, which he called light quanta (and which we refer to as photons).[20] He argued that the amount of power light quanta carry will vary, as Minchin suspected, according to the wavelength of light—the shorter the wavelength, the more power. The shortest wavelengths, for example, contain photons that are about four times as powerful as those of the longest.[21]

Einstein's bold and novel description of light, combined with the discovery of the electron and the ensuing rash of research into its behavior, gave scientists in the second decade of the twentieth century a better understanding of photoelectricity.[22] They saw that the more powerful photons carry enough energy to knock poorly linked electrons from their atomic orbits in materials like selenium. When wires are attached, the liberated electrons flow through them as electricity. While nineteenth-century experimenters called the process photoelectric, by the 1920s scientists referred to the phenomenon as the photovoltaic effect. Solar cells then became a legitimate area for experimentation, reviving the dream shared by Fritts and other nineteenth-century selenium cell researchers—that the world's industries would hum along fuel- and pollution-free, powered by the inexhaustible rays of the sun.[23] Dr. Bruno Lange, a German scientist whose 1931 solar panel resembled Fritts' design, predicted, "In the not distant future, huge plants will employ thousands of these plates to transform sunlight into electric power . . . that can compete with hydroelectric and steam-driven generators in running factories and lighting homes."[24] But Lange's solar battery worked no better than Fritts', converting far less than 1 percent of all incoming sunlight into electricity—hardly enough to justify its use as a power source.

Lange's failure to deliver led experts like E. D. Wilson of Westinghouse Electric's photoelectricity division to judge, "The photovoltaic cell will not even prove interesting to the practical engineer until the efficiency has been increased at least fifty times." Doubting this could ever happen, Wilson declared that the selenium battery "does not show promise as a power converter for solar energy."[25] Dr. Maria Telkes, a chemist at the Massachusetts Institute of Technology and an early advocate of solar technologies, added to Wilson's pessimistic assessment when she reported to a colleague, "The selenium-type photovoltaic cells deteriorate very rapidly when exposed to strong sunlight and the manufacturers claim that it is impossible to prevent this."[26]

Although, the pioneers in photoelectricity failed to build the solar devices that they had hoped to, their efforts were not in vain. One contemporary of Minchin's credited them for their "telescopic imagination [that] beheld the blessed vision of the Sun, no longer pouring unrequited into space, but by means of photo-electric cells . . . [its] powers gathered into electric storehouses to the total extinction of steam engines and the utter repression of smoke."[27] In his 1919 book on solar cells, Thomas Benson complimented their work in selenium as the forerunner of "the inevitable Solar Generator."[28] Telkes also felt encouraged by the selenium legacy, writing, "Personally I believe that photovoltaic cells will be the most efficient converters of solar energy, if a great deal of further research and development work succeeds in improving their characteristics."[29]

## Notes & Comments

1. W. Smith, *The Rise and Extension of Submarine Telegraphy* (1891; reprint, New York: Arno Press, 1974), 310.
2. W. Siemens, "On the Dependence of the Electric Conductivity of Selenium on Light," Monthly Report of the Berlin Academy of Sciences, 17 February 1876, in *Scientific and Technical Papers of Werner von Siemens, 1892–1895,* vol.1 (London: J. Murray, 1892), 175.
3. W. Smith, "The Action of Light on Selenium," *Journal of the Society of Telegraph Engineers* 2 (1873): 32.
4. Ibid.
5. Siemens, "On the Dependence."
6. W. G. Adams and R. E. Day, "The Action of Light on Selenium," *Philosophical Proceedings of the Royal Society of London* 25 (1877): 115.
7. Ibid.
8. W. G. Adams and R. E. Day, 1877, "The Action of Light on Selenium," *Philosophical Transactions of the Royal Society of London* 168 (1877): 341–42.
9. William Grylls Adams, letter to Professor G. G. Stokes, director of the Royal Society, 27 November 1875, Add 7656 RS 1116, Cambridge University Manuscript Collection. As a practical outcome of their work, Adams suggested that properly constructed selenium cells could act as photometers to measure the intensity of light. Indeed, the selenium photometer has enjoyed great commercial success. Almost every camera depends on its selenium light meter to guarantee proper exposure. The selenium cell has also had great commercial success as the "seeing eye" in automatic door openers and as a light-detecting instrument for astronomers. D. Raisbeck, "The Solar Battery," *Scientific American* 193 (December 1955): 102–3.

10. W. Siemens, "On the Electromotive Action of Illuminated Selenium Discovered by Mr. Fritts, of New York," *Van Nostrand's Engineering Magazine* 32 (1885): 392n. Despite the much-deserved praise for Fritts' ground-breaking work, it was Edmond Becquerel, a French experimental physicist, who discovered the photoelectric effect. In 1839, Becquerel reported having observed a current produced when two different metal plates immersed in a liquid were exposed to sunlight. But unlike crystalline selenium, Becquerel's apparatus has no true material affinity to the devices that currently convert sunlight directly into electricity. Siemens also failed to mention Adams and Day.

11. Ibid, 515.

12. Ibid.

13. C. Maxwell, letter to George G. Stokes, 31 October 1876, Archives of the Royal Society, RR. 429-31.

14. Siemens, "On the Electromotive Action," 515.

15. W. Tower et al., *Principles of Physics* (Philadelphia: Blackiston's Son & Company, 1914), 128–29.

16. George Minchin, letter to Sir Oliver Lodge, 28 May 1887, Ms Add 89/172, University College London Library. Minchin expressed similar thoughts in his scientific poem *Naturae Veritas*.

> . . . 'tis true that each successive age
> Brings forth its crowd of theories and rules—
> What yesterday was "law" to-day the sage
> Holds up for scorn or laughter in the schools
> —G. Minchin, 1887

G. Minchin, *Naturae Veritas* (London: MacMillan, 1887), 57.

17. G. Minchin, "Experiments in Photoelectricity," *Proceedings of the Physical Society* 11 (1890): 108–9.

18. G. Minchin, quoted in R. Appleyard, "Photo-Electric Cells," *The Telegraphic Journal and Electric Review* 28 (23 January 1891): 126.

19. Minchin, "Experiments in Photoelectricity," 101–2.

20. According to Einstein's biographer Abraham Pais, "In 1905, Einstein discovered light-quanta without using Planck's law. . . . [T]he physics community at large had received the light-quantum hypothesis [of Einstein's] with disbelief and with skepticism bordering on derision . . . [making him] a man apart in being . . . almost the only one, to take the light quantum seriously. . . . [But] by 1923, [light-quanta were] rapidly accepted by the scientific community." A. Pais, *"Subtle is the Lord. . .": The Science and the Life of Albert Einstein* (New York: Oxford University Press, 1982), 358, 357, 414.

In 1926, the distinguished American chemist Gilbert Lewis gave the name photon to light quantum, the term the scientific community has used ever since. Ibid., 359.

21. D. Chapin, *Energy from the Sun* (New York: Bell Telephone Laboratories, 1962), 24.

22. The electron was discovered in 1897 and quickly won acceptance among physicists. Pais, *"Subtle is the Lord,"* 359.

23. In one of the first texts on photoelectricity, the authors wrote, "the photoelectric effect [of which they considered photovoltaics a subset]. . . now finds itself closely intertwined with recent advances in quantum mechanics and [interest in] the new electron theory . . . has been increasing steadily." A. Hughes and L. Dubridge, *Photoelectric Phenomena* (New York: McGraw-Hill, 1932), v.

24. "Magic Plates Tap Sun for Power," *Popular Science Monthly* 118 (June 1931): 41.

25. E.D. Wilson, "Power from the Sun," *Power* 28 (October 1935): 517.

26. Dr. Maria Telkes, letter to Dr. Dan Trivich, 26 May 1952. (Courtesy Dr. Richard Komp.)

27. R. Appleyard, "Photo-Electric Cells," *The Telegraphic Journal and Electric Review* 28 (23 January 1891): 125.

28. T. Benson, *Selenium Cells* (New York: Spon and Chamberlain, 1919), 60.

29. Dr. Maria Telkes, letter to Dr. Dan Trivich, 31 August 1952. (Courtesy Dr. Richard Komp.)

Chapter Three
# The Dream Becomes Real

The solar cell that ultimately proved interesting to the practical engineer resulted from research into silicon's possible applications in electronics conducted at Bell Laboratories during the early 1950s. Calvin Fuller and Gerald Pearson, two Bell scientists, led the pioneering effort that took the silicon transistor, now the principal electronic component used in all electrical equipment, from theory to working device. Pearson was described by an admiring colleague as the "experimentalist's experimentalist."[1] Fuller, a chemist, learned how to control the introduction of the impurities necessary to transform silicon from a poor to a superior conductor of electricity. As part of the research program, Fuller gave Pearson a piece of silicon containing a small concentration of gallium. The introduction of gallium had given the silicon a number of loosely connected positive charges and the silicon therefore became positively charged. When Pearson dipped the rod into a hot lithium bath, according to Fuller's instructions, it gained poorly bound electrons where the lithium penetrated, and the lithium silicon became negatively charged. Where the positive and negative silicon meet, a permanent electrical force develops: This is the p–n junction. Silicon prepared this way needs but a certain amount of outside energy, which lamplight provided in one of Pearson's experiments, to dislodge weakly linked charges. Inadvertently, he had made a very good solar cell by allowing only a thin layer of negative silicon to cover the positive silicon. Because the p–n junction was extremely shallow, sufficiently energized photons from the light of the lamp were able to free the electrons near the junction. Its built-in electrical force pushed the liberated electrons to metal contacts placed on the silicon and so captured them. Wires connected the contacts to Pearson's ammeter, which, to the scientist's great surprise, recorded a significant electrical current.[2]

While Fuller and Pearson busied themselves improving transistors, another Bell scientist, Darryl Chapin, had begun work on the problem of providing small amounts of intermittent power in remote humid locations. In any other climate, the traditional dry-cell battery would do, but "in the tropics [it] may have too short a life" due to humidity-induced degradation, Chapin explained, "and be gone when fully needed."[3] The Laboratory therefore had Chapin investigate the feasibility of employing alternative sources of freestanding power, including wind machines, thermoelectric generators, and small steam engines. Chapin suggested that the investigation include solar cells and his supervisors approved.

In late February 1953, Chapin commenced his photovoltaics research. Placing a commercial selenium cell in sunlight, he recorded that the cell produced 4.9 watts per square meter. Its efficiency, the percentage of sunlight it could convert into electricity, was a little less than 0.5 percent.[4]

Word of Chapin's solar power studies and dismal results got back

*A drawing by Gerald Pearson of the first silicon solar cell he invented, the forerunner of most solar cells in use today.*

*Gerald Pearson, Darryl Chapin,
and Calvin Fuller (left to right),
the principle developers of the
silicon solar cell, measuring the
electrical energy produced by
one of their cells when exposed
to light.*

to Pearson. The two scientists had known each other for years. They had
attended the same university, and Pearson had even spent time on the
Chapin family tulip farm, so it was only natural that Pearson let Chapin
know about his recent experiment. He advised Chapin, "Don't waste another moment on selenium," and gave him the silicon solar cell that he had
tested.[5]

Chapin's tests, conducted in good strong sunlight, proved Pearson
right.[6] The silicon solar cell had an efficiency of 2.3 percent, about five
times better than the selenium cell. Chapin immediately dropped selenium
research and dedicated his time to improving the silicon solar cell.

His theoretical calculations of the silicon solar cell's potential were
encouraging. An ideal unit, he figured, could use 23 percent of the incoming solar energy to produce electricity.[7] However, Chapin set a realistic
efficiency goal of 5.7 percent before he would consider the cell a viable
power source. Try as he might, after months of hard work Chapin could
not improve the first cell that Pearson had given him.

"The biggest problem," Chapin reported, "appears to be [making]
electrical contact to the silicon."[8] "It would be nice if we could get solder

leads right to the silicon," he explained, "but this is not possible."[9] He therefore had to electroplate a portion of the negative and positive silicon layers in order to tap into the electricity generated by the cell. Unfortunately, no metal plate would adhere very well, thus presenting a seemingly insurmountable obstacle.[10]

Chapin also had to cope with the inherent instability of lithium-bathed solar cells, since lithium migrated through the cell at room temperature. Its transiency moved the original surface location of the p–n junction to deep within the silicon, consequently making it more difficult for sunlight to penetrate the junction where the harvesting of electrons occurs.

Then, an inspired guess changed his tack. "It appears necessary to make our p–n barrier [i.e., junction] very next to the surface [so] that nearly all the photons are effective in delivering a charge to the barrier," Chapin decided. However, his failure to make progress had so shaken his confidence that he felt it necessary "to get some encouragement on [this] new idea."[11] One person Chapin talked with was Russell Ohl, who had initiated research into silicon during the early 1940s. In fact, years before Pearson's discovery, Ohl had tried to turn specially prepared silicon into a solar energy converter, but his cell performed no better than selenium cells.[12] Seeing that Ohl had no answers, Chapin realized that he and his colleagues would have to blaze their own trail.

He therefore turned to Fuller for advice. Two years earlier, while trying to make a transistor, Fuller had made a p–n junction extremely close to the surface—exactly as Chapin had in mind. Instead of making the negative silicon of the transistor from lithium, the chemist explained, he had vaporized a small amount of phosphorous onto the otherwise positive silicon. Fuller said he would make some samples for Chapin.[13] The hope was that "the thin layer may make the short waves from the sun useful," and that the phosphorus coating would prove more permanent than the lithium. Chapin kept his fingers crossed, hoping that the new process might allow for better contacts as well.[14]

The initial optimism quickly faded when Chapin reported a month later, "Nothing exceptional has been produced."[15] Then, acting on a hunch that cell performance was also being hindered by the silicon's shiny surface—which deflected a good deal of sunlight, rather than absorbing and using it—Chapin coated a cell with a clear dull plastic. "With the antireflective covering," Pearson recalled, "[Chapin] got up to 90 percent of the incoming sunlight to reach the [cell]," producing the equivalent of almost forty-one watts per square meter, which greatly surpassed the performance of Pearson's first cell. By converting about 4 percent of the incoming sunlight into electricity, the coated cell edged Chapin closer to his goal of 5.7 percent. Subsequent similarly treated cells continued to better Pearson's original accomplishment. This progress renewed Chapin's hope of creating a silicon solar cell power generator, which called for connecting a number of cells in series. He then dedicated much of his time to this pursuit.[16]

Breaking the 4 percent mark, however, appeared impossible. The lingering failure to achieve good contacts weighed against any further success. While Chapin's work at Bell Laboratories floundered, arch-rival RCA announced that its scientists had come up with a nuclear-powered silicon cell, the atomic battery, coinciding with America's Atoms for Peace program, which promoted the use of nuclear power throughout the world. Instead of sun-supplied photons, it used photons emitted from strontium-90, one of the deadliest residues of radioactive waste, to force the flow of electrons and positive charges near the p–n junction to generate electricity. To showcase its new invention, RCA made a dramatic presentation at Radio City in New York that caught the media's imagination. David Sarnoff, founder and president of RCA, initially famous as the telegraph operator who had tapped out the announcement to the world that the *Titanic* had sunk, hit the keys of an old-fashioned telegraph powered by the atomic battery to send the message "Atoms for Peace."

The atomic battery, according to RCA, would someday power homes, cars, and locomotives with the radioactive waste produced by nuclear reactors.[17] What its public relations people failed to mention, however, was why the venetian blinds had to be closed during Sarnoff's demonstration. Years later one of the lead scientists came clean: If the silicon device had been exposed to the

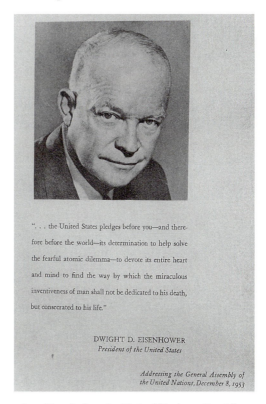

". . . the United States pledges before you—and therefore before the world—its determination to help solve the fearful atomic dilemma—to devote its entire heart and mind to find the way by which the miraculous inventiveness of man shall not be dedicated to his death, but consecrated to his life."

DWIGHT D. EISENHOWER
*President of the United States*

*Addressing the General Assembly of the United Nations, December 8, 1953*

*Speaking before the United Nations, President Dwight D. Eisenhower proposed the use of atomic energy for peaceful purposes. Termed "Atoms for Peace," the program became America's trump card in the Cold War. The country's advocacy of nuclear power effectively counteracted the Soviet portrayal of the United States as a war-mongering nation.*

sun's rays, solar energy would have overpowered the strontium-90. Had the nuclear element been turned off, the battery would have continued to work on solar power alone. The director of RCA Laboratories did not mince words when he ordered his scientists to go along with this deception, telling them, "Who cares about solar energy? Look, what we really have is this radioactive waste converter. That's the big thing that's going to catch the attention of the public, the press, the scientific community."[18]

Right he was! The *New York Times* swallowed the bait, calling Sarnoff's show "prophetic,"[19] and predicting that now the nuclear battery was available, "[T]here is no theoretical reason why even now we should not have hearing aids and wrist watches that run continuously for the whole of a man's useful life."[20]

RCA's publicity coup stirred Bell Laboratories' management to put pressure on the solar cell investigators.[21] Luckily for anxious Bell executives, Fuller came up with an entirely new way to make silicon solar cells that would break Chapin's impasse. He cut silicon into long narrow strips modeled to the dimensions of Chapin's best performing cells. Then he turned the solar cell's configuration on its head. Instead of starting with positively charged silicon, as in all previous experiments, Fuller began by adding a minute quantity of arsenic to give the silicon an excess of electrons. The negatively charged silicon then went into the furnace for a coating of boron—and out came a novel cell. The controlled introduction of boron made the cell's ultrathin face positively charged, with the p–n junction very near the surface.[22] Much to everyone's relief, solving the problem of making good contact with the silicon proved to be a breeze.[23] Several boron samples were also treated with the antireflective coating Chapin had developed. On the first sunny day in early 1954, all three cells did well. One, however, outperformed the rest, reaching an efficiency of nearly 6 percent—the target Chapin had set almost a year before. It was more than fifty times as efficient as the selenium cells of the 1930s, a threshold that engineers of the time, such as E. D. Wilson, felt it necessary to overcome if photovoltaic cells were to be considered for generating electrical power. Chapin was also happy to announce, "[T]he yield of good cells by the boron process is very much better than for the phosphorous process."[24]

Chapin now confidently referred to the silicon cells as "power photocells . . . intended to be primary power sources."[25] Assured of reproducibility and success, Chapin, Fuller, and Pearson then built a number of power photocells and used them to drive a small motor at a press conference.

Proud Bell executives presented the "Bell Solar Battery" to the public on April 25, 1954, displaying a panel of cells that relied solely on sun power to run a 21-inch Ferris wheel. The next day the Bell scientists took their show to the meeting of the National Academy of Sciences in Washington, D.C. There they ran a solar-powered radio transmitter, which broadcast voice and music to the prestigious gathering. The press took notice that "linked together electrically [the Bell solar cells] deliver power from the sun at the rate of fifty watts per square yard [while] the atomic cell announced recently by the Radio Corporation of America delivers a millionth of a watt [over the same area]. Thus the new Bell device deliv-

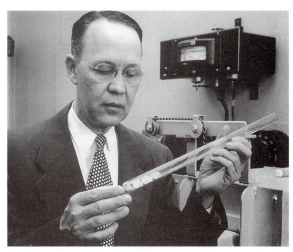

*Calvin Fuller places arsenic-laced silicon into a quartz-tube furnace where he introduced a controlled amount of boron to the material. This resulted in the first solar cell that could generate significant amounts of electricity.*

ers fifty million times more power than the RCA device."[26] A scientist who had worked on the atomic battery lauded the Bell feat as a major technological breakthrough, much like "when aircraft went from propeller speeds to jet velocities."[27] The New York Times concurred, stating on page one that by building the first solar cell that could generate useful amounts of power, the work of Chapin, Fuller, and Pearson "may mark the beginning of a new era, leading eventually to the realization of one of mankind's most cherished dreams—the harnessing of the almost limitless energy of the sun for the uses of civilization."[28]

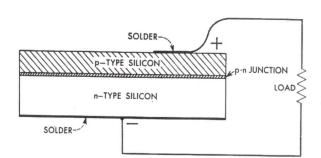

*Cross-section of a Bell Laboratories "power photocell." The depth of the p-type silicon has been enlarged for ease of viewing.*

A Bell scientist, speaking through a microphone powered by silicon solar cells, addresses the audience at the annual meeting of the National Academy of Sciences in 1954, showing the gathering of the nation's best scientists the Laboratories' great advances in solar power.

Bell solar cells powered this toy Ferris wheel. Only one watt of silicon solar cells existed at the time (1954).

Today, a fifty-kilowatt panel drives the Ferris wheel on Santa Monica pier.

## Notes & Comments

1. Interview with Nobel Laureate Dr. Walter Köhn.
2. G. Pearson, 1-20-43 & 3-6-53, Book #23588, Loc. #121-07091, 146–47, 154–55, AT&T Archives, Warren, NJ.
3. D. Chapin, 10-2-52, Book #28161, Loc. 124-11-02, 1, AT&T Archives, Warren, NJ.
4. D. Chapin, 2-12-53, Book #28161, Loc. 124-11-02, 23, AT&T Archives, Warren, NJ.
5. D. Chapin, "Letter to Robert Ford, AT&T Media Relations, Bell Laboratories, Murray Hill, NJ," December 1994. (Courtesy Mrs. Audrey Chapin Svensson); G. Pearson, 3-6-53, Book #23588, Loc. 121-07-01, 155.
6. D. Chapin, 3-23-53, Book #28161, Loc. ibid., 25, AT&T Archives, Warren, NJ.
7. Ibid., 27.
8. D. Chapin, Progress Report for March and April, 6 May 1953, Dept 1730. (Courtesy Mrs. Audrey Chapin Svensson.)
9. D. Chapin, *Energy from the Sun* (New York: Bell Telephone Laboratories, 1962), 56.
10. D. Chapin, Progress Report for July and August, 4 September 1953, Dept. 1730. (Courtesy Mrs. Audrey Chapin Svensson.)
11. D. Chapin, 6-3-53, Book #28161, Loc. #?, 31, AT&T Archives, Warren, NJ.
12. Ohl's best silicon solar cell had a maximum efficiency of 0.725 percent. E. Kingston and R. Ohl, "Photoelectric Properties of Ionically Bombarded Silicon," *The Bell System Technical Journal* 31 (1952): 814.
13. D. Chapin, Progress Report for July and August, 4 September 1953, Dept. 1730. (Courtesy Mrs. Audrey Chapin Svensson.)
14. D. Chapin, 6-17-53, Book #28161, Loc. #?, 31–32, AT&T Archives, Warren, NJ.
15. D. Chapin, Progress Report for May and June, 1 July 1953, Dept. 1730. (Courtesy Mrs. Audrey Chapin Svensson.)
16. G. Pearson, Video Ref #400-07-1105 AT&T Archives, Warren, NJ; D. Chapin, Progress Report for September and October, 1 November 1953, Dept. 1730. (Courtesy Mrs. Audrey Chapin Svensson.)
17. W. Laurence, "RCA Demonstrates Atomic Battery. . . ," *New York Times* (27 January 1954): 1, column 4.
18. Interview with Dr. Joseph Loferski.
19. W. Kaemfert, "First Direct Use of Atomic Energy Puts Up Electrical Current, Small but Prophetic," "Science in Review," *New York Times* (31 January 1954): IV, 9.
20. Editorial, "Nuclear Energy," *New York Times* (27 January 1954): 26.
21. Dr. Morton Prince, assisting Chapin, Fuller, and Pearson, wrote of high-level pressure "to make and assemble a sufficient number of suitable silicon photocells such that the complete package will deliver one watt of power in the sunlight." M. Prince, 2-23-54 & 3-3-54, Book #24651, Loc. ?, 143, 147, AT&T Archives, Warren, NJ.

22. Calvin Fuller had begun experimenting with boron for p-type silicon for photovoltaic cells as early as October 1953. C. Fuller, 10-7-53, Book #24863, Loc. 123-09-01, AT&T Archives, Warren, NJ.

23. D. Chapin, Progress Report for January and February, 1 March 1954, Department 1314. (Courtesy Mrs. Audrey Chapin Svensson.)

24. Ibid.; D. Chapin, 1-26-54 & 2-23-54, Book #29349, Loc. 124-11-02, 30–34, AT&T Archives, Warren, NJ.

25. D. Chapin, "Construction of Power Photocells," Coversheet for Technical Memorandum, 2 March 1954, MM-54-131-9. (Courtesy Mrs. Audrey Chapin Svensson.)

26. "Bell Makes Battery Powered by Sun," *Electrical World* (10 May 1954): 48.

27. Interview with Dr. Joseph Loferski. The late Joe Loferski explained why Bell Labs' silicon solar cell worked so much better when placed in the sun than the silicon atomic battery, which he had helped to produce at RCA Laboratories: "The difference was that the device made at RCA was done by alloying the silicon with the necessary impurities. The important step forward taken at Bell Laboratories was that they made their cells by diffusing the impurities into the silicon. This allowed them to place the p–n junction near the surface, while for the alloyed cell we had the junction near the bottom."

28. "Vast Power is Tapped by Battery Using Sand Ingredient," *New York Time* (26 April 1954): 1.

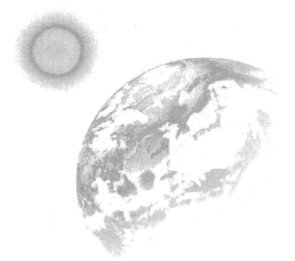

Chapter Four

# Searching for Applications

Few inventions in the history of Bell Laboratories evoked as much media attention and public excitement as the unveiling of the silicon solar cell, referred to as the Bell Solar Battery. Based on details disclosed at the press conference Bell held in April 1954, *U.S. News & World Report* speculated excitedly in an article titled "Fuel Unlimited": "The [silicon] strips may provide more power than all the world's coal, oil and uranium. . . . Engineers are dreaming of silicon-strip powerhouses. The future: limitless."[1]

The Bell invention also gave a boost to the solar energy field. The late John Yellot, a mechanical engineer, who during the 1950s probably knew more about attempts to use the sun's energy than anyone else in America, hailed the silicon solar cell as "the twentieth-century's first really important breakthrough in solar energy technology."[2] In fact, many solar advocates, according to a 1955 *Newsweek* report, foresaw the solar cell's "development as an eventual competitor to atomic power."[3]

Technical progress continued, and in the next eighteen months cell efficiency doubled. But commercial success eluded solar cells because of their prohibitive cost. The price tag of the starting material was in large part responsible. "Unfortunately, the proper properties have not been obtained [for efficient cells] except with very high purity silicon," Chapin explained.

"[With] the present price . . . about \$380 per pound . . . in any discussion of the economics of solar [cells]," he concluded, "this factor stands important." With a one-watt cell costing \$286, Chapin calculated that in 1956 a homeowner would have to pay \$1,430,000 for an array of sufficient size to power the average house. This led him to soberly state, "However exciting the prospect is of using silicon solar converters for power. . . . [c]learly, we have not advanced to where we can compete . . . commercial[ly]."[4]

Chapin's pessimistic analysis did not discourage Western Electric, the Bell subsidiary that transfers developments from the lab to the field. One of the first applications for the Bell solar cell was to help run telephone lines in rural Georgia, where there was no other nearby power source. It did its job well, with one glitch. Bird droppings would foul the panel covers and prevent the sun's rays from reaching the cells. A weekly cleaning solved the problem. However, soon silicon transistors, made in almost the same fashion as silicon solar cells, were adopted to amplify voice messages along rural telephone lines like those in Georgia. Because they needed only minuscule amounts of power to operate, a tiny amount sent through the telephone lines kept everything working, thus making the solar contribution unnecessary.

In 1955, Maurice Paradice, CEO of National Fabricated Products, a company of eighty employees, bought the license to manufacture silicon solar cells from Western Electric to produce modules for the commercial market. The company hired several engineers and physicists to oversee research and development, as well as assemblers to put the arrays together. "The motivation for starting production had been the idea to utilize the no-cost energy from the sun and thus provide cheap power, particularly where power was not readily available, such as in the less developed world," according to Dr. Martin Wolf, the physicist the company assigned to commercialize the silicon

*A lineman installing a Bell solar array to power a rural telephone line in Americus, Georgia.*

solar cell. The firm therefore believed "there should be a very large market available," Wolf added, "and handsome profits should be realizable." However, since the company was unable to bring the price down, "substantial sales did not materialize," Wolf recalled. "Applications for solar cells for all kinds of purposes were [therefore] pursued."[5]

The first major order for solar cells came from the Dahlberg Company, which, Wolf explained, "had been manufacturing . . . eyeglass-mounted hearing aids for some time and was now expanding its product line with solar cell-powered hearing aids, which did not need recharging or replacing the batteries."[6] National Fabricated Products sent Dahlberg a substantial number of small rectangular cells that were mounted in series to the temples of eyeglass frames. Unfortunately for Paradice, Dahlberg suddenly declared bankruptcy without making a single payment.

Hoffman Electronics, of El Monte, California, bought National Fabricated Products in 1956 and zealously tried to sell the idea of solar electricity. Its energetic and innovative CEO Leslie Hoffman showed a gathering of one-hundred and fifty government officials the company's new twenty-five-watt "sun-power to electricity converter-module." "Consider," he told them, "a remote telephone repeater station in the middle of the desert. Power is required but the station is unmanned. An array of solar cells and a system of storage batteries is the answer. Or consider the case of a navigational buoy in the harbor. Here a lamp supplied with solar cells and a small storage battery can run unattended indefinitely. The Forest Service has a severe problem in providing power for its unattended radio relay stations. Either heavy expendable dry batteries or fuel for a rather unreliable gasoline engine charging unit must be 'packed in' to the site at great expense. The solar cell is a much better answer."[7]

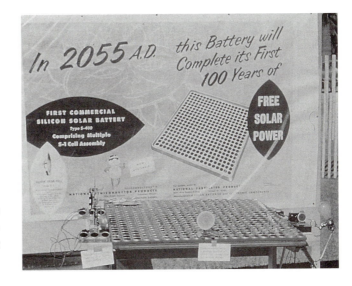

*A National Fabricated Products demonstration, part of the company's attempts to commercialize photovoltaics in the mid-1950s.*

Although he had built and installed prototypes for both the Coast Guard and the Forest Service, no one bought Hoffman's pitch. "The reasons are obvious," William Cherry, an engineer and early solar cell enthusiast, contended in 1955. "Commercial power . . . costs between 1¢ and 2¢ per kilowatt hour. Dry battery power . . . costs $23.70 per kilowatt hour and photovoltaic power . . . would cost about $144 per kilowatt hour."[8]

Desperate to find commercial outlets, novelty items such as toys and transistor radios run by solar cells were manufactured. Wolf recalled a company display "where under room lighting . . . [toy] ships traveled in circles in a children's wading pool. The model of a DC-4 with four electric motors . . . turning the propellers was powered exclusively by the solar cell embedded in the wings."[9] With solar cells running only playthings, the initial enthusiasm generated by the Bell discovery quickly waned. Journalist Harland Manchester, writing in a 1955 issue of *Reader's Digest*, tried to shore up support for solar cells by giving the public a historical perspective. "Viewed in the light of the world's power needs, these gadgets are toys," Manchester admitted. "But so was the first motor built by Michael Faraday over a century ago—and it sired the whole gigantic electrical industry."[10] Despite such consoling words, Darryl Chapin could not help but wonder, "What to do with our new baby?"[11]

*A National Fabricated Products circular shows toys and devices powered by silicon solar cells*

*Zenith produced a limited number of photovoltaic-powered radios. The cost of the solar cells made them very expensive, and they did not sell well.*

## Notes

1. *U.S. News & World Report*, "Sun's Energy: Fuel Unlimited" 36 (7 May 1954): 18.
2. J. Yellot, "Progress Report: Solar Energy in 1960, *Mechanical Engineering* 82 (December 1960): 41.
3. "Power by Sunlight," *Newsweek* 46 (10 October 55): 108.
4. D. Chapin, "The Conversion of Solar to Electrical Power," 18 May 1956, UCLA Lecture Notes, 8. (Courtesy Mrs. Audrey Chapin Svensson.)
5. M. Wolf, "Notes on the Early History of Solar Cell Development Particularly at National Semicondutor and Hoffman Semiconductor Div.," 18 October 1995, 4. (Courtesy Dr. Martin Wolf.)
6. Interview with Dr. Martin Wolf.
7. H. Leslie Hoffman, "Harnessing the Sun's Energy," *Trusts & Estates* 9 (October 1957): 1021.
8. W. Cherry, "Military Considerations for a Photovoltaic Solar Converter," *The International Conference on the Use of Solar Energy—The Scientific Basis, Transactions of the Conference, Electrical Process*, vol. 5 (Tucson, AZ, 31 October–1 November 1955) (Tucson: University of Arizona Press, 1955), 127, 132.
9. Wolf, "Notes on the Early History," 4.
10. H. Manchester, "The Prospects for Solar Power," *Reader's Digest* (June 1957): 73.
11. D. Chapin, "Letter to Robert Ford, AT&T Media Relations, Bell Laboratories, Murray Hill, New Jersey," December 1994. (Courtesy Mrs. Audrey Chapin Svensson.)

## Chapter Five
# Saved by the Space Race

Immediately after hearing of Chapin, Fuller, and Pearson's breakthrough, General James O'Connell, Commander of the U.S. Army Signal Corps, who had a keen eye for technological developments, arranged for the Corps' lead researcher on power devices, Dr. Hans Ziegler, to visit Bell Laboratories. Ziegler, who had worked on Germany's bombing program during World War II and who had come to the United States with Werner von Braun as a valued scientist to help the United States win the Cold War, was truly smitten by what he saw at Bell Laboratories.[1] After his visit, he told his colleagues, "Future development [of the silicon solar cell] may well render it into an important source of electrical power [as] the roofs of all our buildings in cities and towns equipped with solar [cells] would be sufficient to produce this country's entire demand for electrical power."[2] In a briefing to General O'Connell, Ziegler was even more emphatic about the importance of the silicon solar cell, stating, "In the long run, mankind has no choice but to turn to the sun if he wants to survive."[3]

Ziegler and his staff immediately set out to explore "the opportunities of the new invention" for the Signal Corps, whose role was to provide and maintain reliable communication systems for the military.[4] After months of searching, they came up empty-handed, "[e]xcept for one application," Ziegler informed O'Connell.[5] That "one application" was a top-secret opera-

*Dr. Hans Ziegler, the chief advocate for powering satellites with solar cells.*

tion dubbed "Operation Lunch Box": the construction and launching of an artificial satellite. The Signal Corps contended that solar energy was the logical power source for the satellite's communication system and demonstrated that silicon solar cells were best suited for the job.[6] The Air Force took a similar position in a parallel report, proposing that Bell solar cells supply the electricity for any satellite it would launch.[7]

Freed from terrestrial restraints on solar radiation, namely inclement weather and nighttime, "operations above the earth's atmosphere," the Signal Corps believed, would "provide ideal circumstances for solar energy converters."[8] A tiny array could provide the small amount of power that the transistorized communication equipment onboard a satellite required without encumbering the payload. Also, silicon solar cells theoretically would never wear out, unlike the other power option—batteries—which stopped working within seven to fourteen days. The Signal Corps therefore concluded, "For longer periods of operation and limited allowance for weight . . . the photovoltaic principle . . . appears most promising."[9]

Ziegler knew that his idea to use solar cells for powering satellites was not new. "In fact, solar power had long been considered standard equipment in science fiction stories," he wrote.[10] Back in 1945, for example, Arthur C. Clarke advocated the launching of three space stations to provide a worldwide communications network. Improvements in photoelectric devices, Clarke argued, would make it possible to utilize solar energy to provide electricity for running satellite transmitters.[11]

Science fiction came closer to fact on July 30, 1955, when President Eisenhower announced America's plans to put a satellite into space. A drawing that accompanied Eisenhower's front-page statement in the *New York Times* showed silicon solar cells as its power source. The *Times* quoted Dr. S. Fred Singer, the designer, as proposing that "the new solar 'batteries' developed by the Bell Telephone Laboratories be used to generate electrical current for transmission of data from the satellite to earth."[12]

The launch was to be part of an international effort, called the International Geophysical Year (IGY), to implement technologies developed during and after World War II to explore earth and space. To emphasize the cooperative nature of the space portion of the program, Ziegler said, "Every effort was made to [stress] . . . peaceful goals . . . and avoid the impression that it would signal space as the future battlefield and the start of a related weapons race."[13] With this in mind, a special civilian committee made up of leading American scientists reviewed the satellite proposals submitted by the Air Force, the Army, and the Navy. The Navy won the competition. It had suggested that a new rocket be developed for launchings, while the other two branches had proposed the use of projectiles already in service. To the committee, a brand-new rocket seemed less tainted by past militarism and more in line with the purported theme of the international program, the peaceful pursuit of scientific knowledge for the betterment of humanity.

With the selection of the Navy, "the prospects for the introduction of solar power [in space] seemed rather dim," according to Ziegler.[14] From the start, the Navy had specifically ruled out the use of silicon cells as "unconventional and not fully established."[15] With its refusal based on such grounds, Ziegler retorted, the whole space effort should be scrapped since no one had ever attempted to launch a satellite, either. Ziegler and his staff's consistent efforts "to convince the Navy that solar cells could be hardly less dependable than limited-life chemical batteries received little response."[16]

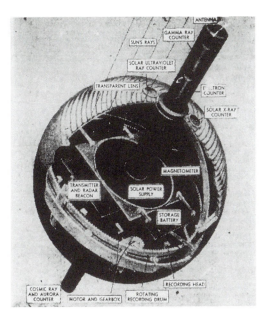

*The announcement, on the front page of the* New York Times, *that the United States intended to launch a satellite into space was accompanied by this illustration, which shows that the satellite's designer planned to use solar modules to power its communication equipment.*

The Navy's stubborn refusal to yield to reason drove Ziegler to conduct a crusade to put solar cells on America's first satellite. His burning zeal "to give mankind the benefit of this invention at the earliest possible time"[17] overrode any qualms about violating protocol. He therefore took his case to another forum, the Technical Panel on the Earth Satellite Program, a civilian group of prominent academic scientists who oversaw the development of America's fledgling space program. The Signal Corps presented the panel with a "Proposal for the Use of Solar Power during the IGY Satellite Program," offering "advice and assistance to all interested internal instrumentation projects."[18]

Unlike the cold shoulder Ziegler had received from the Navy, the civilian scientific panel embraced his ideas enthusiastically. The panel did not like the idea of relying on batteries either, because they automatically condemned "most of the on-board apparatus . . . [to] an active life of only a few weeks [while] nearly all of the experiments will have enormously greater value if they can be kept operating for several months or more." Hence, the scientists sided with Ziegler, saying, "It is of utmost importance to have a solar battery system" onboard.[19]

Relenting to pressure from the civilian oversight committee, the Naval Research Laboratory was forced to ask the Signal Corps to participate in its satellite program, Project Vanguard, and assigned it the responsibility of designing a solar cell power system.

Ziegler and his staff readily developed a prototype that clustered individual solar cells onto the surface of the satellite's shell. They designed the modules to provide "mechanical rigidity against shock and vibration and to comply with thermal requirements of space travel."[20] To test their reliability in space, the Signal Corps, in cooperation with the Naval Research

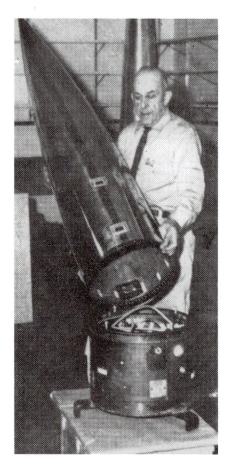

*A Navy technician attaches a nose cone embedded with solar cells to a rocket that will be launched to test whether the cells can survive the rigors of space.*

Laboratory, attached cell clusters to the nose cones of two high-altitude rockets. One rocket reached an altitude of 126 miles (203 kilometers), the second 192 miles (310 kilometers), both high enough to experience the vicissitudes of space. "In both firings, the solar cells operated perfectly," Ziegler reported to an international conference on space activity held in the fall of 1957.[21] A U.S. Signal Corps press release added, "The[ir] power was sufficient for satellite instruments . . . [and they were] not affected by the temperatures of skin friction as the rockets passed through the atmosphere at more than a mile a second."[22] Still, skepticism persisted. The *New York Times* revealed its doubts when reporting the launches. In a rare case of editorializing in a news story, the words "proof" or "proved" were placed within quotation marks when telling of the Army's analysis of the test results. The article also indicated that the Navy had not dropped its opposition to solar cells, despite their success in a space environment, reporting, "At least the first four satellites probably will have conventional chemical batteries as their power source."[23]

The break for Ziegler and his staff came in August 1957. The Vanguard program had become mired in problems and delays. To fast-track a launch, it was decided to put a number of grapefruit-sized spheres into orbit containing nothing but a transmitter.[24] The altered plans, according to Ziegler, presented "a splendid new opportunity to give our solar power supplies a free ride," since "a considerable weight capacity would remain free. The Navy, after much pressure, yielded and allowed this unused weight to be utilized by a cluster of solar cells and its transmitter" on the first satellite set for launching.[25]

The Signal Corps immediately contacted Hoffman Electronics, the firm that had tried unsuccessfully to interest the government in buying its solar modules for land-based purposes. Eugene Ralph, a Hoffman engineer at the time, recalls assembling some of the cells for the Vanguard.

After several failed attempts, the first satellite with solar cells aboard went into orbit on St. Patrick's Day, 1958. Nineteen days later, a *New York Times* headline read: "Vanguard Radio Fails to Report/Chemical Battery Believed Exhausted/Solar Unit Functioning."[26] Celebrating the first anniversary of the Vanguard launch, the Signal Corps let the public know that "[t]he sun-powered Vanguard I . . . is still faithfully sending its radio message back to earth."[27]

The small Vanguard satellite equipped with solar cells proved far more valuable to science than the first two, and much larger, Sputniks,

*A Hoffman Electronics scientist holds the Vanguard that initiated solar power's use in space. The scientist is pointing to one of the four solar cell clusters that the company manufactured.*

whose reliance on conventional batteries silenced them after a week or so in space. "[T]he longevity of [the Vanguard's] solar-powered transistor radio outshines them all," the Signal Corps boasted.[28] The long-lived transmitter allowed mapmakers to pinpoint the locations of Pacific islands and enabled geophysicists to better determine the earth's shape.

Ziegler accurately assessed the legacy of Vanguard for photovoltaics, writing two years after the launch, "This solar supply. . . has broken down the prejudice which existed at the time toward the use of solar cells in space applications."[29] More important, without solar energy, Ziegler pointed out almost twenty years later, "not much of our past or future exploration and economical practical application of space would have been possible."[30]

Ziegler later advocated the construction and launching of solar-powered communication satellites with the same zeal he had shown for solarizing the Vanguard. At first, opposition to his new fight appeared as insurmountable as it had in the Vanguard controversy. Several decades later, Darryl Chapin noted that Dr. Ziegler must again have felt vindicated "with modern worldwide instant communication by satellite as a regular experience."[31] In the spring of 1998, the world learned the importance society places on solar-run communication satellites when the loss of one such satellite caused 90 percent of America's forty-five million pagers to fall silent.

Ziegler and Chapin admired each other's work and corresponded on several occasions in the late 1970s. Although Ziegler's name had slipped from the memory of an aging Darryl Chapin fifteen years later, the co-discoverer of the silicon solar cell continued to appreciate the importance of Ziegler's contribution. "I wish I could remember the name of the scientist [who] engineered to put solar cells on the little grapefruit [satellites]," he wrote, "because he ranks among the great and should be remembered."[32]

## Notes & Comments

1. After his arrival in the United States, Ziegler "became the key civilian in charge of [the Signal Corps' space efforts] and he represented the Army and the Signal Corps in related high-level national and international conferences [as well as having been] appointed by the National Academy of Sciences as U.S. Delegate to the IGY [International Geophysical Year] Conference in Moscow." Brigadier General H. McD. Brown, "Part II—Score and Beyond," Army Communicator (Winter 1982): 17.

   Ziegler became Chief Scientist at the U.S. Army Signal Research and Development Laboratory in 1959, and gained a reputation as one of the world's leading authorities on satellite instrumentation and communications.

2. H. Ziegler, "News Bulletin to the Staff at Ft. Monmouth Signal Corps," 11 May 1954 & 18 May 1954. (Courtesy Dr. Hans Ziegler.)

3. H. Ziegler, "Utilization of Solar Energy." Briefing to Lt. Gen. J. O'Connell, September 1955, 2, 6, 7. (Courtesy Dr. Hans Ziegler.)

4. H. Ziegler, letter to D. Chapin, 1979. (Courtesy Dr. Hans Ziegler.)

5. Ziegler, "Utilization of Solar Energy," 9.

6. H. Ziegler, "A Signal Corps Odyssey, Part 1—Prelude to Score," *Army Communicator* (Fall 1981): 23.

7. J. D. Wartham, "Information on Dr. S. E. Singer's Proposal on *Minimum Orbital Unmanned Satellite, Earth*." 14 September 1954, Air Force Materiel Command, Wright Patterson Air Force Base, HQ-AFMC/HO.

8. W. Cherry, "Military Considerations for a Photovoltaic Solar Energy Converter," *The International Conference on the Use of Solar Energy—The Scientific Basis, Transactions of the Conference, Electrical Process*, vol. 5 (Tucson, AZ, 31 October–1 November 1955) (Tucson: University of Arizona Press, 1955), 128.

9. Signal Corps Engineering Laboratories, *Proposals for Satellite Program* 1 (9 September 1955): 25–26. (Courtesy Dr. Hans Ziegler.)

10. Ziegler, "A Signal Corps Odyssey."

11. A. C. Clarke, "Extra-Terrestrial Relays," *Wireless World* 51 (October 1945): 306–7.

12. "U.S. to Launch History's First Man-Made Earth-Circling Satellite," *New York Times* (30 July 1955): 7.

13. Ziegler, "A Signal Corps Odyssey," 19.

14. H. Ziegler, letter to D. Chapin, 1979. (Courtesy Dr. Hans Ziegler.)

15. H. Ziegler, personal communication, 1996. (Courtesy Dr. Hans Ziegler.)

16. H. Ziegler, letter to D. Chapin, 1979. (Courtesy Dr. Hans Ziegler.)

17. Ibid.

18. J. van Allen, "Minutes of the Third Meeting of Working Group on Internal Instrumentation of the I.G.Y. Technical Panel on Earth Satellite Program," 11

October 1956, Iowa City, Iowa. James van Allen Papers, Box 249, Folder 7, The University of Iowa Archives, Iowa City, IA.

19. Report of the Working Group on Internal Instrumentation of the USNC/IGY Technical Panel on the Earth Satellite Program. Presented at the Conference on Use of Solar Batteries for Powering Satellite Apparatus, Washington, D.C., 30 April 1955. James van Allen Papers, Box 251, Folder 7, The University of Iowa Archives, Iowa City, IA.

20. H. Ziegler, "Solar-Power Sources for Satellite Applications," *Annals of the International Geophysical Year* (1958) pt. 1–5, 3.3.3., 302–3.

21. H. Ziegler, 'Solar Power Sources for Satellite Applications." Paper presented at the International Geophysical Year Rocket and Science Conference, Washington D.C., 30 September–5 October 1957. (Courtesy Dr. Hans Ziegler.)

22. U.S. Army Signal Garrison, Public and Technical Information Division. Press Release, 18 June 1957. ECOM/Historical Research Collection, Ft. Monmouth, NJ.

23. "U.S. Army Finds Solar Batteries Ideal for Operating Satellite Instruments," *New York Times* (21 June 1957): 3.

24. Ziegler, "A Signal Corps Odyssey," 23.

25. Dr. Hans Ziegler, personal communication, 1996. (Courtesy Dr. Hans Ziegler.)

26. "Radio Fails as Chemical Battery Is Exhausted: Solar-Powered Radio Still Functions," *New York Times* (6 April 1958): 37.

27. Office of Technical Liaison, Office of the Chief Signal Corps Office, Department of the Army, "Sun-Powered Vanguard I Satellite Marks First Anniversary on March 17, 1959," Washington, D.C., 1. ECOM/Historial Research Collection, Ft. Monmouth, NJ.

28. Ibid., 2.

29. H. Ziegler, "The Signal Corps at the Space Frontier," *IRE Transactions on Military Electronics*, MIL-4, #4 (October 1960): 405.

30. H. Ziegler, letter to D. Chapin, 1979. (Courtesy Dr. Hans Ziegler.)

31. D. Chapin, letter to to H. Ziegler, 1979.

32. D. Chapin, "Letter to Robert Ford, AT&T Media Relations, Bell Laboratories, Murray Hill, New Jersey," December 1994. (Courtesy Mrs. Audrey Chapin Svensson.)

Chapter Six
# Bringing Solar Cells
# Down to Earth

Sputnik III, launched three weeks after the first Vanguard, was the
first Russian satellite to have its telemetry system powered by silicon
solar cells. It appears that Russia's premier space scientists had taken a
keen interest in Chapin, Fuller, and Pearson's work, which had become
well-known throughout the world, and put what they learned to good
use. Their success with solar cells on Sputnik III led Russian space scientist
Yeogeniy Fedorov to predict in the summer of 1958, "Solar batteries
. . . would ultimately become the main source of power in space."[1] Events
proved Fedorov right.

Despite their success on the Vanguard and Sputnik III, many in the
space business still considered solar cells a stopgap measure, a technology
to use until nuclear power could be developed. "A lot of people were
worried that solar would not be powerful enough for the larger space
probes that were in the works," said John Goldsmith, who began his photovoltaics career at General Electric's missile and space division in 1960.
However, "nuclear technologies," according to Goldsmith, "were never
able to deliver the performance, the reliability, and the safety that people
had earlier anticipated."[2] In contrast, the pessimism held toward solar's
power limitations and reliability proved wrong. Engineers and scientists
working with satellites came to accept the solar cell as "one of the critically

important devices in the space program," since it "turned out to provide the only practical power source in space [at a reasonable distance to] the sun."[3] By 1972, about one thousand American and Soviet spacecraft relied on solar cells for electricity, while fewer than ten of the six hundred American satellites ran on atomic batteries. Power supplied by solar cells had increased from milliwatts on Vanguard I to many kilowatts on Skylab A, generating the electricity to transmit and receive data; to operate computers, scientific instruments, and stabilizing equipment; and to maintain proper temperature controls.[4]

The urgent demand for solar cells above the earth opened an unexpected and relatively large business for the companies manufacturing them. "On their own commercially, they wouldn't have gotten anyplace," observed the late Dr. Joseph Loferski, who spent his life working in photovoltaics.[5] But locked into the space race with the Russians, the American government poured more than fifty million dollars into solar cell research and development from 1958 to 1969.[6] "For the first time in the long history of solar energy research," the late John Yellot observed, "relatively large amounts of government funds are being assigned to projects which will lead to the building of . . . reliable solar power devices."[7] Indeed, as Martin Wolf contends, "The onset of the Space Age [was] the salvation of the solar cell industry."[8]

The astronomical price of silicon solar cells, however, kept their use confined to space. Despite a price drop of about 300 percent between 1956 and 1971, solar cells still cost $100 per watt, two hundred times the price of electricity in those years.[9] It was, in fact, the very exacting requirements of the space industry that forced the cost of solar cells to remain so inordinately high.

*This Hoffman Electronics illustration from the 1960s shows a variety of spacecraft powered by silicon solar cells.*

Space cells had to be extremely over-engineered to withstand bombardment from high energy particles and micro-meteorites. There was no room for failure. If for any reason the cells did not work, the mission was lost and millions of dollars of equipment rendered useless. In the 1960s and 1970s, no one could be sent up to make repairs. Cell efficiency, rather than the price of the energy they delivered, dominated their design. The more power per pound an engineer could pack into a module, the lighter the payload. This reduction in weight reduced the size of the engine required for liftoff and that, in turn, saved

*The need for solar cells created by the space race created a new industry.*

a lot of money. Also, each satellite required a different module design, because each mission had a different power requirement. This sporadic demand for customized products discouraged the mass production essential for bringing prices down.

While things were looking up for solar cells in space in the late 1960s and the very early 1970s, down on earth electricity from the sun seemed as distant as ever—with one exception. Government agencies engaged in covert activities immediately realized the value of solar cells to help them in their intrigues. The CIA, for example, wanted to know the extent of traffic on the Ho Chi Minh Trail during the Vietnam War. Circumstances obviously ruled out the use of on-site human counters. Instead, the organization employed the Special Forces to install large camouflaged arrays of photovoltaics to run clandestine heat-sensitive metering devices.[10] Like the space program, covert organizations don't worry about cost. According to one enterprising solar cell salesman, who had set up shop right down the street from the Pentagon, "[Government] people would walk in with briefcases full of money and they'd walk out with solar modules. For what they had to do, money was no object."[11]

For applications on earth where people did have to consider the economics, Dr. Harry Tabor, one of the most respected solar scientists of the time, painted a bleak picture. "Most experts seem to agree," Tabor wrote, "that silicon cells as we know them today are not going to get much cheaper."[12] Tabor's dismal assessment seemed right on target. In a working paper prepared for the 1973 international solar summit, "The Sun in the Service of Mankind," its author lamented, "[Although] space research in particular has led to impressive progress in . . . photovoltaic . . . conversion, it has not led to parallel developments from the point of view of the use of solar energy on earth."[13] Yet, a month later, at the conference, researchers from Exxon (then Esso) surprised everyone by announcing that its subsidiary Solar Power Corporation "has recently commercialized and is currently marketing a [photovoltaic] module . . . which will compete with other power sources for . . . earth applications."[14]

Solar Power Corporation, the company that finally brought solar electricity down to earth, was founded and run by Dr. Elliot Berman, an industrial chemist, with Exxon providing the financing.[15] Berman's solar quest began in 1968 after he had completed a ten-year assignment which put his former employer, Itek Corporation, into the photographic materials business. When his boss, Dick Philbrick, asked what he wanted to do next, Berman replied that he didn't know. The company president then made him a generous offer: "I'll pay you to think about it." After months of deliberation, Berman concluded that he wanted to do something that would have important societal impact. Discovering "a close relationship between energy availability and quality of life," he decided to investigate better ways to provide electrical power, especially to those most in need, those who lived in rural areas of developing countries. After much research, it appeared to Berman that "[i]n the long term . . . only nuclear fusion would meet [future] needs and this simple-minded chemist chose the easier course, using our existing fusion plant—the sun."[16]

When his employers asked if they should be interested, Berman's honest reply was that he wasn't sure. So, he was given six months to prepare a proposal for the company's scientific board. Titled "Solar Power," it boldly stated, "This document outlines a program for the utilization of solar energy to meet the world's power needs." Conventional silicon cells would not be considered, it said, because of "their extremely high cost."[17] Instead, a completely different cell would be researched, developed, and marketed, a solar cell made like the photographic film with which he had previously worked. This process, Berman believed, would lead to a major reduction in the technology's cost.

When the company rejected Berman's solar dream, "I picked up my marbles and left. I told my wife it might take six months to find some money for my solar project." Venture capitalists, however, scoffed at the idea. "They weren't very venturesome and what I had wasn't a venture by their definition," Berman recalled.[18] Eighteen months had passed when a chance conversation with a director on the board of Berman's former employer, a gentleman who worked for the Rockefellers, led to Exxon's doorstep. The timing could not have been better. Exxon had just formed a task force to look thirty years ahead to determine where the energy business was going. The group had concluded that energy prices would be considerably higher in 2000 and that these higher prices would create opportunities for alternative sources of energy, if their costs dropped sufficiently. Solar energy, particularly photovoltaics, seemed the most likely candidate. Solar cells' limitless life, modularity, and capability to generate electricity cleanly and directly from an inexhaustible power source without bulky ancillary equipment intrigued the Exxon investigators. When the Exxon people heard Berman's proposal to make a very low-cost terrestrial photovoltaic device, "We found it very exciting. We conceived the end product as being very cheap, being used as simple roofing material," the group leader recalled.[19] Exxon immediately invited Berman to join its laboratory in Linden, New Jersey, in late 1969 to pursue his research.

Berman also convinced Exxon that it would be valuable to commercialize an interim product while research on the film concept proceeded. This way the company could gauge the market for cells, determine the needs of potential customers, and learn how solar cells would fare under everyday conditions. As Berman pointed out, "People who sit in the lab and design products . . . end up many times with things that are not really workable because it's hard to define what the real world needs [while] locked in an ivory tower."[20]

To keep their feet on the ground, Exxon and its solar team conducted surveys to determine where the markets were—if, indeed, there were markets. At the 1970 price of $100 per watt, they learned there would be few takers. If the price dropped to $20 per watt, the survey revealed a sizable demand, because then solar cells would be cheaper than nonrechargeable primary batteries or thermoelectric generators to power equipment beyond the reach of utility lines.

Berman did not want to dilute his long-term research efforts by building modules for the commercial venture, so he and his staff searched the world for a reasonably priced solar cell. Except for some offers of reject space cells at cut-rate prices, no one could provide them. (Buying reject cells did not interest the Exxon researchers because their needs greatly exceeded the entire space-cell supply.) The problem, Berman discovered, was that everyone in the space-cell business was hung up on efficiency, which is just not an issue on earth. The primary criteria for any terrestrial power device, Berman knew, "is how many kilowatt-hours you get for a dollar."[21]

Left no choice but to make its own, Berman's group followed the space industry and started with crystalline silicon wafers. But there the similarity with space endeavors ended. When Berman and his colleagues addressed the challenge of making cheaper cells, they found that no major breakthroughs were necessary to bring the price down. For example, Solar Power Corporation did not start with very expensive pure semiconductor-grade crystalline silicon, as had the space industry. Instead it bought the much cheaper silicon wafers rejected by the rest of the ever-growing semiconductor industry, but which were perfectly suited for generating power. For satellite use, wafers were cut from a cylindrical crystal and trimmed into rectangles. This made for a more compact fit, but it wasted a lot of expensive silicon in the process. (This waste was incidental to the space program, though, in light of the savings realized by reduced area and weight.) But for earth applications, limiting the size of the modules was not as compelling as conserving the expensive crystalline silicon. Berman therefore used crystalline silicon wafers as they were prepared for use in the semiconductor industry.

In space-cell manufacture, after the silicon stock was sawn into wafers, the rough surfaces were polished and coated with an antireflectant so that the cell would absorb, not reflect, sunlight. What had not been realized was that "if you take a wafer as sawn," Berman explained, "it has a nice, dull matte finish which was perfect for our needs."[22] This discovery meant that two production steps could be eliminated without affecting the cells' intake of sunlight. Another change in the manufacturing process was to increase the size of the cell, which reduced the number of cells needed per module and, consequently, the number of connections between them. Berman also introduced less expensive materials for packaging the modules, such as acrylic plastic for the panel cover, silicone rubber in which to embed the cells, and a circuit board onto which the cells were soldered. He was the first to recognize that silicon solar cells could tolerate imperfections that other semiconductor devices could not. Discoveries such as these cut costs in a big way. By early 1973, the scientists at Solar

*Dr. Elliot Berman testing various solar arrays manufactured by his new company, Solar Power Corporation, for terrestrial uses.*

Power Corporation were making single-crystal silicon modules for $10 per watt and selling large quantities for around $20 per watt, bringing to earth what hitherto had been principally a space-based enterprise.[23]

## Notes & Comments

1. "Reds Boast of Solar Battery," *Electronics*, Business Edition 31 (11 July 1958): 33.
2. Interview with John Goldsmith. NASA bureaucrats had so little confidence that solar cells could power large future satellite projects that they spent more money on other approaches, such as nuclear, which never panned out, than was spent on photovoltaics. Interview with Dr. Martin Wolf.
3. National Research Council, Ad Hoc Panel on Solar Cell Efficiency, *Solar Cells Outlook for Improved Efficiency* (Washington, D.C.: NRC, 1972), 3; M. Wolf, "Notes on the Early History of Solar Cell Development Particularly at National Semiconductor and Hoffman Semiconductor Division" (18 October 1995), 7.
4. National Research Council, Ad Hoc Panel on Solar Cell Efficiency, *Solar Cells Outlook for Improved Efficiency* (Washington, D.C: NRC, 1972), 3, 5.
5. Interview with Dr. Joseph Loferski.
6. National Research Council, *Solar Cells Outlook*, 3, 67.
7. J. Yellot, "Solar Energy in 1960," *Mechanical Engineering* 82 (December 1960): 41.
8. Wolf, "Notes on the Early History," 7.
9. D. Chapin, "The Conversion of Solar to Electrical Energy," May 18, 1956, UCLA Lecture Notes, 18. (Courtesy Mrs. Audrey Chapin Svensson); National Research Council, *Solar Cells Outlook*, 21.
10. Interview with Peter Iles.
11. Interview with Arthur Rudin.

12. H. Tabor, "Power for Remote Areas," *International Science and Technology* (May 1967): 54.

13. A. Moumouni, "Small- and Medium-Scale Applications of Solar Energy and Their Potential for Developing Countries" (Paris: UNESCO, 1973), 15.

14. B. Kelly et al., "Investigations of Photovoltaic Applications," *Photovoltaic Power and Its Applications in Space and on Earth* (Bretigny-sur-Orge, France: Centre National D'Etudes Spatiales, 1973), 511. In this volume of papers on photovoltaics presented at the UNESCO Solar Summit held in 1973 in Paris, Joe Lindmayer, the founder of Solarex, lists his affiliation as working for COMSAT. Dr. Wolfgang Palz, a leading European photovoltaics researcher, remembers that at dinner one night during the conference Joe Lindmayer was mulling over whether to start his own terrestrial photovoltaics company. Interview with Dr. Wolfgang Palz.

15. As a young engineer in the 1970s, Bernard McNelis, now head of IT Power, one of the world's leading photovoltaics consulting firms for developing countries, wanted to enter the solar industry. He researched the field thoroughly and discovered that "Solar Power Corporation was the driving force that brought a space technology to earth." Interview with Bernard McNelis.

    Dr. Richard Blieden, who headed the National Science Foundation's task force on solar energy in the early 1970s, agrees with McNelis' assessment. The written record also confirms Solar Power's priority in the terrestrial photovoltaics market. "Solar Cell Prices Dropping, But Broad Terrestrial Use Awaits Mass Production," *Electronics* (19 July 1973): 40.

    Other pioneering companies, such as Solarex and Solar Technology International, followed Solar Power into the terrestrial market several years later. These two companies have evolved into the largest photovoltaics companies in the world.

16. E. Berman, quoted in "Founders of PV: Elliot Berman," *Photovoltaics International* IV, no.3 (April/May 1986): 31.

17. E. Berman, "Solar Power." Proposal prepared for Itek Corporation, 1968.

18. Interview with Elliot Berman.

19. Interview with Lou Shrier, Exxon Corporation.

20. Interview with Elliot Berman.

21. Ibid.

22. Ibid.

23. "Sunlight Converted to Electricity," Boston *Sunday Globe* (19 August 1973): reprint.

Chapter Seven
# The First Mass Earth Market

Once the price of solar cells was reduced to a competitive level for earth use, marketing began in earnest. Although some people talked about acres of solar cells supplying electricity to America's homes and industries, those at Solar Power Corporation did not share that vision.[1] With modules producing electricity for around forty times the price Americans were paying for electricity generated at power plants, no one at the company expected to compete with the utilities. "It was never even in our thinking," asserted Clive Capps, international sales representative for Solar Power Corporation in the early 1970s.[2] "Most of the applications where it is economically sensible to use these modules are in remote locations, where it is difficult to run a wire," the company declared in 1973.[3]

For example, Solar Power considered powering navigation aids, such as flashing lights and foghorns on buoys, with solar cells since electrical wires rarely came close to where they were moored. In fact, Elliot Berman assumed that "the big buyer would be the Coast Guard."[4] Some years before, in an effort to branch out from the space market to the terrestrial, Hoffman Electronics had installed a prototype system for the Coast Guard. The Coast Guard engineer and aids-to-navigation specialist in charge of the experiment, Maurice Lostler, expressed his belief that "[o]ne day Coast Guard lights, horns, whistles and other navigational aids will be powered

*Elliot Berman next to a module that is powering a navigation aid in Tokyo Harbor. Notice the wire spikes positioned to discourage seagulls from perching.*

by that old celestial fireball, the sun."[5] However, because solar radiation data for the entire country had not yet been collected, the Coast Guard felt it could not properly design systems for regions as diverse as Maine, Washington State, the southern California Coast, and Florida.[6] It therefore did not press forward with powering its navigation aids with photovoltaics.

While in Japan in the late 1960s, on his quest to find a relatively inexpensive solar cell, Berman took a trip to Tokyo Bay to see a solar-powered buoy that had been in operation for some time.[7] He saw firsthand that solar cells could effectively run navigation aids when the battery terminals were well-protected from corroding salt water, and "the sea-gull issue" resolved. Since cleaning the panels after each bird visit was out of the question, something had to be done to discourage perching. The Japanese found that ringing the array of panels with wire spikes did the job.

Back in America, with renewed confidence, Berman told the Coast Guard about the Japanese success, hoping that the agency would give solar cells a second chance. Although the principal decisionmaker turned out to be a high-school chum, he proved to be "the stumbling block." He wanted to do more research, but, as Berman laconically stated, "You can do research forever."[8]

With the Coast Guard out of the picture for the time being, someone at Solar Power suggested running the navigation lights and horns on Exxon's oil platforms in the Gulf of Mexico with photovoltaics. In the Exxon offices in New Jersey, where Solar Power Corporation was also headquartered, the idea was dismissed as sheer folly. "Those platforms are loaded with power! Why would they need solar energy?" Exxon executives exclaimed.[9]

Rather than accept this opinion, Berman and his colleagues headed to the Gulf to check out the situation. The fact-finding trip revealed that while, yes, the platforms where the crews lived had plenty of power, the adjacent working platforms (which constituted the majority) had none. They also learned that when Exxon put the first platforms in the Gulf, in the late 1940s, the Coast Guard had told the company to install lights because without them the platforms were hazards to navigation. Being an offshore novice at the time, Exxon didn't have the slightest notion of where to buy the lights. Thumbing through the Houston phone book, an Exxon troubleshooter stopped at the listing "Lighthouse, Incorporated" and dialed the number. A fellow named Nathan "Available" Jones answered; he ran a garage in East Houston where he installed two-way radios in cabs and air horns and lights on ambulances.

The Exxon representative explained the problem and asked Available if he could take care of it. Although he had no more than an eighth-grade education, Jones, an enterprising man who had come to Houston with only a quarter in his pocket, could hear opportunity knocking. "No problem," he said. "By the way, who told you that you have to have lights on those platforms?" Upon hearing it was the Coast Guard, he immediately called the agency to find out where to purchase the lights. When he learned that the sole manufacturer of navigation lights in the United States was located in New Jersey, Available was on the next flight east. He soon became the company's distributor for the Gulf.[10]

By and by, oil companies were making strikes all over the Gulf, inundating it with platforms. Not only did they buy lights from Available Jones, they also bought

*Nathan "Available" Jones, founder of Automatic Power.*

huge nonrechargeable batteries, which were much like large flashlight batteries weighing five hundred pounds or more, to power them. Available's new company, Automatic Power, developed into a multimillion-dollar business that made most of its money installing, servicing, and replacing those batteries, which wore out in a year or less.

And because the battery business became the source of much of Jones' growing wealth—keeping him awash in Rolls Royces and attractive young women—he was not about to let a better technology dead end his road to riches. He therefore closely monitored the development of energy sources that might threaten his livelihood.

About twelve years into the business, in 1959, Jones learned of Hoffman Electronics' attempt to commercialize photovoltaic-powered navigation aids. Soon thereafter Automatic Power bought the prototype and shelved it, preventing anyone on the Gulf from getting wind of what it judged to be a superior technology. Guy Priestley, who succeeded Available Jones as president of Automatic Power, made no bones about the company's motivation for suppressing photovoltaics. "We didn't want to bring solar on the market because we were selling primary batteries. And we knew that once we brought the solar panels out we weren't going to sell any more batteries. . . . So we said, 'S__t! This is going to cut into our market. Let's not put these things out until we have to.'"[11] Competition eventually forced Available's hand.

The battery business had become so frenetic that Automatic Power was swamped. "Salesmen quit answering the phone by midweek," according to current president Steven Trenchard. "They couldn't make deliveries and there was nothing but irate customers on the line."[12] So some enterprising salesmen from Automatic Power decided to start their own company, Tideland Signal Corporation. Undercapitalized as it was, the new firm found it nearly impossible to break Automatic Power's hold on the market.

Through its research, Solar Power learned that Automatic Power had about 70 percent of the business and Tideland Signal held the rest.[13] "Tideland was definitely struggling," Berman recalled, so naturally, Solar Power approached Automatic Power first.[14] The people at Automatic Power candidly told the Solar Power sales representative that "what you're talking about doing is going to decrease our sales," and they showed him out. His presentation received an entirely different reception at Tideland. The executives there embraced Solar Power's product, convinced that "here is the chance to increase our market share."[15]

Tideland really pushed solar modules and offshore platform owners soon realized that this new product would save them a lot of money. Reliability, paramount for safety equipment, made biweekly servicing and frequent replacement of nonrechargeable batteries a necessity. Moving them on and off the platforms was a chore: The batteries were heavy and highly toxic. All the boat trips and helicopter rides out to the platforms to tend the batteries, to bring in new ones, and to take old ones back to shore made for a very steep bill, not to mention the high cost of the batteries themselves. In contrast, when a sun-charged battery went bad, a replacement cost $160, compared to $2,100 for a nonrechargeable battery. Furthermore, the entire photovoltaic-powered system could be transported by a small standby skiff. Moving a nonrechargeable battery called for a crane boat at $3,500 per day.[16] With such advantages, the oil and gas industry rapidly took to photovoltaics. "It saved us time and money, and that, of course, is better," stated a veteran navigation aids engineer. By the mid- to late 1970s, hundreds of modules had been sold for use on the ever-increasing number of oil platforms. All the major oil companies—Amoco, ARCO, Chevron, Exxon, Texaco, and Shell—were buying. Automatic Power, watching its market share slip into Tideland's hands, had to come out with a solar product, too. The company bought modules from another photovoltaics pioneer, Bill Yerkes, who in the late 1970s and early 1980s would head the world's largest photovoltaics company, ARCO Solar (now Siemens Solar).

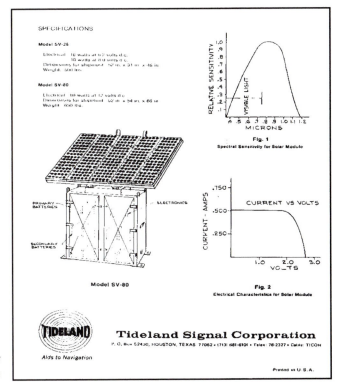

*Tideland Signal's first sales brochure featuring Solar Power Corporation modules.*

Technological improvements in navigation aids made solar even more economical. New types of lenses and sound equipment required less power, which reduced the number of panels needed to run them. The market for solar-powered navigation aids got another major boost in late 1978, when the Environmental Protection Agency (EPA) outlawed dumping batteries into the ocean. Prior to EPA's action, crews servicing platforms often threw the highly toxic batteries overboard instead of ferrying them to port for disposal. The public was outraged when the evening news showed mountains of used batteries, which contained mercury among other toxic substances, sticking out of the water near the platforms at low tide. This led to a federal crackdown. Once the EPA strictly enforced the proper handling and disposal of these hazardous items, no one could afford to use huge batteries offshore.

By 1980, just seven years after their introduction, solar-powered navigation aids had become standard fare in the Gulf of Mexico. That year, a reporter from *Forbes* hitched a ride on a shrimp boat to view solar's penetration of the Gulf market. Out at sea, "amid hundreds of oil platforms, stretching as far" as he could see, the reporter took out his binoculars and saw that each platform held "a rectangular panel about four foot by two foot with silvery circles encased in it."[17] Those, of course, were photovoltaic panels. As it turned out, the twenty to twenty-five thousand lights that photovoltaics powers in the Gulf dwarf any business that the U.S. Coast Guard could have offered. "If you just got a piece of that market," Priestley stated, "you did pretty well."[18] And because both Tideland Signal and Automatic Power conducted business in more than one hundred countries, photovoltaic-powered navigation aids spread across the world.

Offshore was not the only place where the oil and gas industry needed power but had difficulty accessing it. For example, much of the Hugoton field, a huge gas-producing area centered in southwestern Kansas and extending into Oklahoma and Texas, had no power source nearby. As one engineer recalled, "You were out in the middle of nowhere."[19] The wells passed through the Glorietta Formation, which contains salt water that quickly corrodes well casings, making it impossible to extract the gas. Had there been utility lines close by, a corrosion-proofing solution would have been simple: Send current into the ground adjacent to the casings to electrochemically destroy the offending molecules that cause corrosion. But in remote areas, such as most of the Hugoton field, "costs prohibit bringing in conventional electrical power lines."[20]

*Securing a photovoltaic array on an oil platform in the Gulf of Mexico.*

*Photovoltaics powers navigation aids on gas and oil platforms throughout the Gulf of Mexico.*

At such sites, corrosion specialists have commonly resorted to placing a dissimilar metal next to the item needing protection. The two differently charged metals act as opposite poles of a battery, which causes a current to flow into the earth. At the Hugoton field, however, not enough power could be generated in this way to keep the well casings corrosion free. So thermoelectric generators, fueled by natural gas or propane, were brought in. The engineers overseeing corrosion protection at the Glorietta Formation wells had hoped to run the thermoelectrics with locally produced gas, but, unfortunately, its high sulfur content destroyed the generators' flues. The gas also contained a lot of moisture, which froze in winter. What's more, the generators suffered corrosion problems of their own.

The failure of alternative power sources to effectively protect well casings came to the attention of Larry Beil, a corrosion specialist assigned to the Hugoton field. An article Beil had read in a popular science magazine gave him the idea to try solar cells. With photovoltaics, Beil thought, that powerful southwestern sun could generate the electricity needed to keep the wellheads intact. At Beil's suggestion, his company bought some space-reject cells from Spectrolab, a company that built modules for satellites. Though the cells had failed to pass the rigorous standards set by the space program, they worked just fine on earth. And with interest in space winding down in the early 1970s, cell manufacturers, who had relied on satellites for their bread and butter, needed to pursue new markets. However, the effort required to wire all those very small space cells into a panel made for a costly product. In an attempt to bring down the price, Beil experimented with concentrating the sun's energy onto the array. He thought that increasing the amount of sunlight on each module would cut the number of modules needed at each site and thus reduce the overall cost. But this did not happen. Each cell had to be placed at the end of a polished metal funnel and the entire module had to follow the sun throughout the day to keep the concentrating devices in focus. Adding concentrators and tracking mechanisms to the module proved more expensive than using the original Spectrolab product. Beil's pursuit of an economical photovoltaic system led him to Solar Power's factory in Massachusetts. "Our costs drove us to use their modules," Beil admitted.[21]

*This advertisement from the early 1970s introduces solar cells as a power source for corrosion (cathodic) protection.*

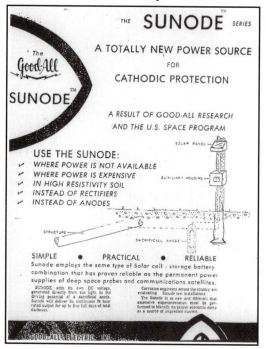

Having at last found a cost-effective product to work with, Beil took it into the field and developed a surefire method for selling clients on solar. "We put in a trial system for one company. It worked, the company was happy," he explained. "Then more and more were bought until [the gas] field was full of them." Presently, Beil estimates, "There are probably between four and five thousand photovoltaic systems" providing electrical current to protect underground wells and pipelines in the United States.[22]

Even more photovoltaic anticorrosion systems have been installed outside the United States, as gas and oil fields are developed worldwide to meet rising energy demands. In most of these locales, such as North Africa and the Middle East, sunlight is abundant and access to utility-generated electricity is limited. For example, the Saudi Arabian–Mediterranean pipeline runs through extremely desolate, sun-drenched territory. "You're talking about thousands of miles of pipe needing protection yet distant from any source of power," Lou Shrier, an Exxon executive, explained.[23] As a consequence, the use of photovoltaics for protecting wellheads and pipelines kicked into high gear, and by 1980 corrosion protection was an important market.

Photovoltaics has also helped petroleum companies in their search for new oil fields. In the mid-1970s, Shell Canada Resources began relying on solar panels to operate surveying equipment that obtains readings from satellites. For precise positioning, the readings have to be taken over a three- to four-day period. Because photovoltaics allowed the surveying equipment to work automatically, rather than manually as had previous equipment, Shell did not have to fly technicians in daily to evaluate the data, a very costly proposition.[24]

Oil and gas producers also need to confirm that the electricity is indeed protecting the casings from corrosion. In addition, they must keep track of the amount of oil and gas flow, and at what pressures and temperatures, from rigs and wells in faraway places. Meter readers could be sent to monitor this data, driving hundreds of miles a day or, if offshore, taking a boat or a helicopter. Instead, the industry has chosen to use photovoltaics, not only to run remote field data acquisition devices, but also to power the transmission of that information to the main office.[25]

"The oil and gas industry got us started," observed Bill Yerkes. "We had the solution for problems they faced. And they had the money to purchase the solar equipment that was needed."[26] This, according to Yerkes, "helped keep us alive during our first years in the business."

*Solar electricity protects this wellhead in Kansas from underground corrosion.*

As Steve Trenchard of Automatic Power remarked, "The commercialization of photovoltaics would have been delayed by at least a decade if not for the oil industry."[27]

The oil companies' firsthand experience with photovoltaics convinced them of the technology's potential. With skyrocketing oil and gas prices in the late 1970s and early 1980s, many in the petroleum business could see that photovoltaics might one day evolve into a major energy source. Almost all the major oil firms have invested heavily in solar cells, buying controlling interests in many fledgling photovoltaics companies. People distrustful of big business believe there is some sort of a conspiracy at work, that the oil companies will buy up and then crush potential competitors. Realistically though, as Charlie Gay, former research director at ARCO Solar and now president of the photovoltaics firm ASE Americas, pointed out, "You don't spend hundreds of millions of dollars in research and development as the oil companies have to destroy a technology."[28]

## Notes & Comments

1. National Research Council, Ad Hoc Panel on Solar Cell Efficiency, *Solar Cells Outlook for Improved Efficiency* (Washington, D.C.: NRC, 1972), 21–22. Government agencies, such as the National Science Foundation (NSF) and Sandia National Laboratories, pushed funding for the development of large centralized photovoltaic electrical generating plants that would produce from one hundred to seventeen hundred megawatts. H. Richard Blieden, "The Status of the National Science Foundation Photovoltaic Program." Paper presented at the Annual Meeting of the International Solar Energy Society, United States Section (Fort Collins, CO, 20–23 August 1974); D. Shueler and B. Marshall, "Photovoltaic System Performance, Applications, and Prospects for Economic Viability in Central Utility Networks," *American Nuclear Society Transactions*, Alternative Energy Power & Systems, Solar Energy—Potential for Large-Scale Production—1, paper #3, 26 (1977): 1.

   The emphasis on "large central stations to produce solar electricity" relegated photovoltaics to "some distant future" when the price of solar cells would dramatically fall, *Science* magazine complained. A. Hammond and W. Metz, "Solar Energy Research: Making Solar after the Nuclear Model?" *Science* 197 (15 July 1979): 241.

   In contrast to such "pie in the sky" research, Solar Power Corporation searched for cost-effective applications. From the beginning, Berman and his colleagues pursued real markets on every continent, establishing photovoltaics beachheads throughout the world in just a few years. This became the basis for the terrestrial photovoltaics industry. In the process, "Elliot [Berman] planted the seeds for a lot of the photovoltaic products we see today," affirms Arthur Rudin, now manager of product marketing for Siemens Solar, one of the largest manufacturers of solar modules in the world. Interview with Arthur Rudin.

2. Interview with Clive Capps.

3. P. Kelly et al., "Investigation of Photovoltaic Applications," *Photovoltaic Power and Its Applications in Space and on Earth* (Bretigny-sur-Orge, France: Centre National d'Etudes Spatiales, 1973), 514.

4. Interview with Elliot Berman.

5. "Energy of Sun Utilized at Los Angeles Harbor," *Hoffman Transmitter* 25, no. 3 (March 1959): 6.

6. Only after the publication of George Löf and colleagues' *World Distribution of Solar Radiation* by the University of Wisconsin in 1966 did such data become readily available. G. Löf et al., *World Distribution of Solar Radiation*, Report #21, College of Engineering, The University of Wisconsin Engineering Experiment Station Report, Madison, WI, 1966.

7. The Japanese pioneered the use of photovoltaics for navigation aids. Sharp Corporation, for example, concentrated on supplying photovoltaic panels to power lighthouses. From 1961 through 1972, Sharp solarized 256 lighthouses situated along the Japanese coastline. Y. Tarui, "Japanese Photovoltaic Systems," *Japanese/United States Symposium on Solar Energy Systems, Summaries of Technical Presentations,* vol. 2 (Washington, D.C., 3–5 June 1974) (Washington, D.C.: MITRE Corp., Washington Operations, 1974?), 7-1. However, Sharp and other Japanese photovoltaics manufacturers outfitted navigation devices with expensive solar cells diverted from their space industry. The companies did not initiate engineering solutions to establish a cost-competitive photovoltaic product for earth applications, as had Solar Power Corporation.

8. Interview with Elliot Berman.

9. Ibid.

10. Interview with Guy Priestley.

11. Ibid.

12. Interview with Steve Trenchard.

13. A U.S. Department of Energy survey of the international photovoltaics market in the late 1970s confirmed that "Tideland Signal Corporation and Penwalt Corporation [parent company of Automatic Power at the time] are the leading developers of PVs for navigational aids." "Assessment of the Technology Transfer of Federal Power Systems, Final Report," June 1979, ALO-4261-T8, 42.

14. Interview with Elliot Berman.

15. Ibid. Tideland Signal later manufactured its own modules.

16. D. Nolan, "Solar Energy Used for Production Applications," *The Oil & Gas Journal* 76 (6 March 1978): 82.

17. J. Briggs, "Solar Power—For Real," *Forbes* 126 (13 October 1980): 142.

18. Interview with Guy Priestley.

19. Interview with Larry Beil.

20. "Solar Energy Tapped for Cathodic Protection of Casing," *Oil and Gas Journal* 78 (20 October 1980): 113.

21. Interview with Larry Beil.

22. Ibid. Corrosion control has given the photovoltaics industry a lot of business.

23. Interview with Lou Shrier.

24. "Solar-Powered Surveyors," *Sunworld* 3, no. 4 (1979): 105

25. The use of photovoltaics in the field has had other benefits. Eliminating the need for utility-generated electricity has also eliminated squabbles over powerline right-of-ways, and so has improved relations between utility companies and landowners.

26. Interview with Bill Yerkes. Sales figures from the late 1970s for solar modules purchased commercially in the United States prove Yerkes' point. Oil companies bought 70 percent of all modules sold on the American market. M. Eckhart, "Assessment of Photovoltaic Industry Markets and Technologies," 30 September 1978, TID-29 300, 111–4.

27. Interview with Steve Trenchard.

28. Interview with Charlie Gay. Some people have feared that the oil companies would buy up all the photovoltaics patents and then quash the technology. But the patent to the silicon solar cell had run out long before the oil companies started to invest heavily in photovoltaics. Fritz Wald, who worked with Tyco Laboratories, a producer of photovoltaics bought out by Mobil and later known as Mobil Solar, called the conspiracy theory "ridiculous." If the oil companies wished to destroy photovoltaics, he said, "they could have much more easily let the technology drop by not investing in it at all." Interview with Fritz Wald.

    The #4 bunker oil burned by power plants, which photovoltaics would replace, "is really the lowest profit commodity for any oil company," Wald explained. "By entering the photovoltaics business from that end, Mobil wasn't worried that photovoltaics would replace" such a low-profit item. "[Photovoltaics] therefore wasn't really trespassing on Mobil's turf at all." Interview with Fritz Wald. Instead, the oil companies that bought into or bought out photovoltaics firms "were really determined to make solar a business," stated Lou Shrier, an Exxon executive who oversaw Solar Power Corporation. Interview with Lou Shrier.

Bill Yerkes, another photovoltaics pioneer, concurs with Wald's assessment and adds that the large rise in oil prices in the fall of 1973 put a lot of extra cash into the pockets of the major oil companies. They were looking for investments and determined that buying photovoltaics companies would be a wise hedge against the future when oil again becomes scarce. But, contrary to their expectations, the price of oil dramatically dropped in the 1980s and it continues to fall, forcing some oil companies to retreat to their core business. Faced with an oil glut, it appeared that solar cells would not become a major player in the power market in the time frame the companies had established. They therefore got rid of their photovoltaics subsidiaries. But many firms took great care in finding new owners. ARCO, which had bought out Yerkes' start-up photovoltaics firm, Solar Technology International, "didn't just close the door" when it decided to discontinue its involvement in photovoltaics, according to Yerkes. "In fact, they pursued selling [it] for a couple of years beyond when any prudent person might have just dumped it." ARCO finally sold it to Siemens, a company they thought would continue to pursue photovoltaics. Likewise, Mobil found a German buyer interested in expanding its innovative photovoltaics technology. Amoco remains in the photovoltaics business, as do British Petroleum (BP) and Shell, who, over the last several years, have significantly increased their financial investments in the commercializing solar cells. In fact, immediately after BP took over Amoco, the company infused Amoco's solar unit, Solarex, with around $50 million, making it the largest photovoltaics company in the world, as of spring 1999.

## Chapter Eight

# Captain Lomer's Saga

Switching from nonrechargeable batteries to photovoltaics made even more economic sense for the Coast Guard than for the oil industry. Its larger buoys measured thirty-eight feet (twelve meters) from top to bottom and their hulls contained huge pockets filled with batteries. Whenever the batteries went bad, a Coast Guard officer explained, a ship called a buoy tender had to be sent out "to handle a buoy, to bring it up on deck to remove the batteries," costing the agency around a thousand dollars an hour. Hence, the price of maintaining batteries far exceeded the buoys' original cost, which made the search for a power source of higher "reliability and longer life absolutely essential," in the estimation of Steve Trenchard, who began his career with the Coast Guard. But because the federal agency was insulated from the forces of competition, "photovoltaics came on faster" on the oil platforms, Trenchard affirmed. "Instead, we studied the dickens out of it. We might have studied it for another decade if it weren't for Captain Lomer."[1]

Lloyd Lomer graduated from the Coast Guard Academy and did postgraduate work in physics and optics. While working on navigation aids in 1973, he became interested in photovoltaics. On his own initiative, Lomer procured some surplus cells. (He and John Goldsmith, who was working as a solar engineer at Jet Propulsion Laboratory in the 1970s, "did a little horse trading for the good of the government" to get them.) Lomer put the cells through tests at the Coast Guard Laboratories even though he had no formal approval to conduct such work, finding that he and sympathetic colleagues "could combine our work so our superiors wouldn't get angry at us." The results convinced Lomer that solar cells were "just what the Coast Guard needed." His superior officers, however, were less enthusiastic and consistently turned down requests for an officially funded solar cell hardware procurement program.

Their rejections did not shake Lomer's confidence or his perseverance: "I simply knew that photovoltaics for the Coast Guard was going to happen sooner or later." His belief in the technology drove Lomer to persist despite the possibility of jeopardizing his career.

The only technical barrier to success that Lomer could foresee was the need to protect the cells and their numerous electrical connections from the sea's constant wave action and corroding salt. Solar panels on buoys require more protection than those on oil platforms because they are closer to the water. But making panels impregnable to the high seas did not appear to be an insurmountable problem. According to Lomer, "If you didn't pinch your pennies, you could seal them up from anything that nature could throw at them."

Having specialized in ocean engineering, and being responsible for improving all of America's navigation aid systems, Lomer was aware of the costliness of large batteries and their maintenance. He therefore knew from the start "that you could spend an awful lot on these solar cells and [still] get your money back really quickly."

Lomer eventually won approval to install a solar unit on a buoy in Ketchikan, Alaska. Placing it in a sun-starved location that receives over 180 inches (4.6 meters) of rainfall per year was a calculated high-stakes gamble. If it worked up there, photovoltaics would do well anywhere the Coast Guard had navigation aids. "The risk paid off," Lomer recalled. Its successful operation in such a challenging location ratcheted up interest in solar cells within the Coast Guard.

All evidence seemed to support the then-Lieutenant Commander. Lomer confidently approached his commanding officer with a proposal for formal funding to begin a conversion program from primary batteries to photovoltaics—but he still could not win him over. The rejection was, in Lomer's opinion, "one of the toughest things that happened in my service career. I had the data, the information, yet I couldn't get my funding proposals past my boss's desk and that was the bottom line."

This time, however, he had another place to turn for help. As the energy crisis of the 1970s unfolded, the Energy Research and Development Agency (ERDA), which later became the Department of Energy, began to fund demonstration projects to show the American people the practical side of alternative energy technologies. Unlike the Coast Guard, the Department of Energy welcomed his proposal. "If we didn't have the best application for solar cells, I would like to know who ever would," Lomer recalled. Those in charge of funding encouraged him to develop a formal plan. Unfortunately, that plan required the signature of his commanding officer. Lomer vividly remembers the meeting: "I went into his office and said, 'Look, boss, you've been telling me for years there's no money for this. But now I have the problem solved.' He looks at me surprised and says, 'What do you mean?' Then I explained about the interest the Department of Energy had. 'No, we can't do that,' [he said], shaking his head

*Captain Lloyd Lomer (Ret.), the crusading Coast Guard officer who almost single-handedly engineered the Guard's conversion from non-rechargeable batteries to photovoltaics as the primary power source for its equipment.*

indignantly. 'We're the Coast Guard, an independent agency, and we're not going to go begging to any other government agency.'"

Lomer then did something drastic, which he refuses to divulge: "Let's just say one day I got the OK to put in a proposal for over a million or so dollars to the Department of Energy, which I already knew was going to be favorably [acted upon]." Very soon thereafter, by 1977, the courageous Coast Guard officer had a funded project and a new boss whom Lomer described as "wonderful." "He said not only are we going to do this, but we are going to get us a team" dedicated to the project.

Solarizing the Coast Guard's installations then began in earnest, with Lomer as the project manager. In his new position, according to Admiral Yost, the Commander of the Coast Guard at the time, "Lomer drafted a plan

THE WHITE HOUSE
WASHINGTON

September 29, 1986

Dear Captain Lomer:

At the beginning of this Administration, I promised the American people that a major priority would be to make Government work efficiently and economically.

Because Federal career employees play a key role in determining how well Government operates, I pledged in August of 1984 to give special recognition to those whose efforts above and beyond the requirements of the job had brought about significant savings -- enabling Government to function better at less cost to the taxpayer. I am delighted to be able to fulfill that pledge by commending and honoring you for your exemplary service and your outstanding accomplishments.

In saving a substantial amount of the taxpayers' money through your initiative and managerial effectiveness as project manager for the conversion of aids to navigation from battery to solar photovoltaic power, you have set an outstanding example for others by demonstrating that Federal employees can make the critical difference. You've also exemplified the Coast Guard motto, Semper Paratus -- Always Ready.

You have my deep appreciation for a job well done.

Sincerely,

Ronald Reagan

Captain Lloyd R. Lomer, USCG
Commandant (G-CMA)
2100 Second Street, S.W.
Washington, D.C. 20593

*A letter from President Ronald Reagan praising Captain Lomer for his pioneering efforts which lead the Coast Guard to switch to solar power.*

to consider all aspects including . . . new hardware procurement . . . site selection, reports, management, training programs, designs, logistic support, and implementation planning."[2] To make doubly sure that the hardware was up to the task, Lomer devised an accelerated testing facility known as "the pit." It earned its formidable name. Here modules were submerged in salt water, subjected to extreme pressure, and cycled through hot and cold water. Only a very rugged, well-protected module with a glass cover and rigid backing could survive. The Coast Guard then stipulated the high standards to which modules had to conform to the bidders in the photovoltaics industry. The competition for the multimillion-dollar contract was fierce. Solarex, one of the fledgling manufacturers of solar cells, was awarded the job.

By the 1980s, the Coast Guard had decided to convert all of its navigation aids to photovoltaic power. Admiral Yost credited Lomer for the change, stating that Lomer was "the driving force of the Coast Guard's conversion of aids to navigation to solar photovoltaic power, and has become the Service's foremost expert on this power source."[3] President Ronald Reagan also commended Captain Lomer "for your exemplary service and your outstanding accomplishment in saving a substantial amount of the taxpayers' money through your initiative and managerial effectiveness as the project manager for the conversion of aids to navigation from batteries to solar photovoltaic power."[4] Despite his Commander-in-Chief's salutation for a job well done, and several medals from the Coast Guard for his pioneering work in solar cells, "Lomer's persistence for photovoltaics cost him his promotion to Admiral," said Steve Trenchard, who had served under Lomer. "It leads you to conclude that sometimes there is no justice in the world."[5]

Captain Lomer, now retired from the Coast Guard and living in Florida, can see the results of his crusade each time he sails his aptly named boat—*Don Quijote*.

Photovoltaic-powered navigation aids are now in world-wide use. Prior to converting to solar electricity, most countries, other than the United States and Canada, ran marine warning lights with acetylene gas fueling an open flame contained in a lanternlike enclosure. A clock-regulated mechanism determined when the gas would ignite. The capital and maintenance costs for such equipment were enormous, even more expensive and troublesome than primary batteries.

The French Lighthouse Service started replacing acetylene systems with solar modules in 1981, making the choice primarily on economic grounds. Analysts reported that "the turnover time of the investment [for photovoltaics] is less than one year."[6] Greece found the economics for replacing acetylene-powered lighthouse lights with photovoltaics compelling as well. In 1983, a gas-powered unit cost $15,300, while the initial investment for its photovoltaic equivalent was only $2,000. Maintenance costs were reduced, too. By 1983, Greece had installed photovoltaics in most of its 960 lighthouses and buoys. Greek solar cell experts called the program "the most successful [photovoltaic] application at present in Greece."[7] In fact, throughout the world "marine applications were precedent-setting in many people's minds," John Goldsmith asserts.[8] Proving itself in the harshest environments was an excellent measure of the technology's potential. People took notice and began to consider photovoltaics for a range of tamer terrestrial uses.

*Solarex modules power this buoy's warning light and horn.*

*Today, solar cells power the lights in almost every lighthouse run by the Coast Guard.*

## Notes & Comments

All quotations and information not otherwise attributed were taken from interviews and correspondence with Captain Lloyd Lomer (Ret.).

1. Interview with Steve Trenchard.
2. A. Yost, "Citation to Accompany the Award of the Coast Guard Achievement Medal to Commander Lloyd R. Lomer, United States Coast Guard," 8 April 1981.
3. Ibid.
4. President Ronald Reagan, letter to Captain Lomer, 29 September 1986.
5. Interview with Steve Trenchard.
6. C. Guy, "Experience of the French Lighthouse Service in Powering Aids to Navigation either on Land or Flotting [sic] Aids with Photovoltaic Generators," *Proceedings of the 2nd International Photovoltaic Science and Engineering Conference* (Beijing, China) (Hong Kong: Printed by Adfield Advertising Company, 1986), 670, 673.
7. K. Soras and V. Makios, "Feasible Stand-Alone Photovoltaic Systems in Greece," *Fifth E.C. Photovoltaic Solar Energy Conference* (Athens, Greece) (Dordrecht: Kluwer Academic Publishers, 1984), 487.
8. Interview with John Goldsmith.

## Chapter Nine
# Working on the Railroad

Solar Power Corporation initiated its plan to market solar cells on earth around the time the world's space programs were winding down after the completion of the United States' Apollo program, which was capped by the historic moon walk. Those associated with the space race began to worry about their professional futures. To ameliorate the gloomy prospects, NASA entered the terrestrial photovoltaics field, approaching it as another mission, like the moon shot. It hoped to recycle its engineers and thus keep its workers employed.[1] NASA believed that a large government appropriation would solve everything.

For those unaccustomed to the NASA–aerospace culture, "This approach was an eye opener," Dr. Allan Rothwarf confessed. He remembers that the principal presentations at one NASA-sponsored conference were given by an engineer from Texas Instruments and a scientist from RCA. Both gave an analysis of "what it would take to make photovoltaics economically viable" for supplying electrical power to America's homes, businesses, and factories. According to Rothwarf, the engineer argued that "a subsidy of $1 billion was needed between 1973 and the year 2000, preferably given to a single company—Texas Instruments—so that they could learn how to make photovoltaics economically. . . . The RCA scientist's presentation was pretty similar, though he only needed half a billion!"[2]

In sharp contrast to the desire for the government to underwrite the American photovoltaics industry, Elliot Berman, who testified before Congress at that time, countered, "There is no need for federal support for the business. . . . [A]t the present time, there exists a commercially viable business based on utilizing silicon photovoltaic devices to provide power in places on earth where sunlight is available and other forms of energy expensive." As an example, Berman informed Congress that "solar energy is [already] being successfully converted to electricity [to] . . . power . . . navigation warning lights and horns on unmanned offshore platforms and maritime buoys worldwide." If Congress really wanted to help the photovoltaics industry grow commercially, Berman suggested that the federal government make policy changes that would benefit the American people rather than provide handouts to the industry. For example, he urged Congress to require warning lights and guards wherever a railroad track and a roadway intersect. Such a policy, Berman argued, would both save lives and promote photovoltaics because many of the 175,000 unprotected crossings in the United States at the time were located far from any source of electricity.[3]

Though Berman's words did not persuade the government to act, in 1974 a salesman at Solar Power, who had previously been employed in the railroad industry, convinced the higher-ups at Southern Railway to try an experimental panel to power a crossing signal near Rex, Georgia. Not convinced that these newfangled wafers could power much of anything, railroad workers connected the solar array to a utility line for backup. "All of us were a little bit skeptical of the technology," stated Bob Mitchell, who has worked for the Southern for many years.[4] But, a funny thing happened in Rex, Georgia, that turned quite a few heads. On several occasions that winter, ice buildup caused the wires to fall. And the only elec-

*A train passes by the first solar-powered crossing signal, Rex, Georgia.*

tricity for miles around came from the solar array. "Just the reverse of what they [had] expected happened," chortled Arthur Rudin, who had installed the panels and who periodically checked the installation. "That's what sold them on the technology."[5] Or as Bob Mitchell stated, "Rex, Georgia, taught the Southern that solar worked."[6]

While the solar experiment continued to purr along at Rex, the Southern Railway found itself on the horns of a dilemma at the Lake Pontchartrain trestle near New Orleans. Colored signal lights had been recently installed, and they required more current. Bringing in utility power was out of the question. "It's just swamp out there for miles," according to Bob Mitchell.[7] To sink the poles into the lake would have cost a fortune, and they could not be placed on the adjacent levee because flood control regulations would not permit it. Instead the Southern tried a thermoelectric generator powered by LP gas. Unfortunately, it attracted swarms of mosquitoes from the swamp reeds at dusk. They would fly over the flue, singe their wings, and fall into the generator. Enough mosquito bodies accumulated to clog the engine's air intakes.

No amount of spraying could keep the mosquitoes away, so the railroad gave up on thermal generators and tried nonrechargeable primary batteries. Along the Southern, the track maintenance crews built their own from a recipe that called for water and sulfuric acid poured into a jar that contained a lead plate. Each signal required thirty-two jar-batteries, which together weighed over six hundred pounds. The homemade batteries worked, but they were a high-maintenance item. On Lake Pontchartrain, the batteries cost more to make and maintain than elsewhere on the Southern because fresh water had to be carted out to the signal site; the lake's salt water would not do. So the Southern Railway was forced to search for another power source.

*Modules along the trestle spanning Lake Pontchartrain ensure that the traffic signals function.*

"Our experience with solar power at Rex, Georgia, had been good," commented the late J.T. Hudson, an executive in the Southern's Communications and Signals Division, "so we decided to give solar a try."[8] That decision was made in 1975 and "the panels are still working," Bob Mitchell attests, who was at the site when they were first installed. The only worry is salt spray buildup. If the salt coating gets too thick, the panels don't charge the accompanying batteries well. "You get a little soap and water and wash the panels and rinse them off. It's that simple," Mitchell said. "Then they're good for another two or three months."[9]

The success at Lake Pontchartrain increased the Southern's confidence in the technology and the railroad expanded its use, especially into track circuitry. Analogous to air traffic control, the objective of track circuitry is to keep trains at a reasonable distance from one another and so prevent head-on collisions or back-enders. The presence of a train changes the rate of an electric current that runs through the track. A decoder picks up that change, translates it, and then throws signals and switches up and down the track to ensure safe passage for all trains in the vicinity. Prior to 1976, the Southern used its nonrechargeable jar-batteries for track circuitry, too. But not only did the labor-intensive technology make these batteries increasingly costly, new government rules, like those regulating battery disposal offshore, upped the ante. As Bob Mitchell explained, "Until the environmentalists got strong, we could just throw the sulfuric acid on the side of the tracks. But you better not be caught doing that anymore! We can't even dispose of the batteries ourselves. We have to have somebody come in and do it for us."[10] Just as in the case of offshore platforms, environmentally sound procedures prescribed by law brought photovoltaics to the fore.

Another government mandate put photovoltaics on the rooftops of some of the Southern's cabooses. A federal regulation adopted in 1976 stipulated rear-end lighting for trains. Long-range cabooses already carried enough power to easily comply, but those on local runs did not. The railway first tried to generate the needed electricity by attaching a fan belt to the rear axle of the caboose. As the train went down the track, the fan belt would rotate and produce electricity. But the belt fell off all the time, and the constant vibrating tore the generator system apart. Again, the Southern turned to photovoltaics. By early 1979, twelve experimental solar installations were in service. They worked so well that the Southern converted all eighty of its local-service cabooses to photovoltaic power. "So far as we know," re-

*Solar cells mounted on the roof of this caboose powered its government-mandated back warning light.*

marked John F. Norris, the Southern's General Superintendent of Communications and Signals, "this is the first use of solar on trains."[11]

Ironically, despite the success in Rex, Georgia, the Southern Railway has not put solar to work powering warning devices at its other unprotected grade crossings in remote locations. That is because the federal government has yet to require them. Nonetheless, some grade crossings in the United States are photovoltaic-powered, such as along the Burlington Northern Santa Fe's track in eastern Arizona and near Phoenix. "Obviously, this area is a prime candidate for any solar application" because of the year-round abundance of sunshine, James Le Vere, manager of special projects for the railroad, attested. To bring utility power to these locations "would have cost hundreds of thousands of dollars. In addition, the number of people using the roads" that cross the tracks "is very high," Le Vere added, but the railroad "traffic is significantly less than we would expect on a normal main line." Hence, the crossing signals along this line operate just a few times each day, which assures that the solar electricity stored in the batteries will always be sufficient. In general, though, especially in the more northerly latitudes, railroads have shied away from using photovoltaics at crossings. "Should the batteries fail and you don't know about it, you're in deep trouble," Le Vere stated. "You're talking about the motoring public and law suits. The liability is just too high."[12]

Solar power, though, has helped railroads free themselves from the burden of utility poles, which require constant vigilance and maintenance. In the early days, telegraph lines were usually installed at the same time that the railroad tracks were put in. Messages sent through the wires by telegraph and later by telephone kept stations informed of arrivals, delays, track conditions, and other matters paramount to the safe and smooth functioning of a railway. By the mid-1970s, wireless microwave technology could do the same, rendering telecommunications through wires obsolete. The Kansas City Southern led the movement to uproot this antiquated technology because the power lines along its tracks in Louisiana and Arkansas had proved particularly difficult to maintain. The long growing season in that region made extra work for the maintenance crews. "We had to fight vegetation, we had to fight pole rot, we had to fight the damage done by hurricanes," complained Stanley Taylor, communications and signaling engineer for the Kansas City Southern. "We had to fight everything that goes with a lot of sun and rain."[13]

Taylor admitted that his company would have preferred that an electric utility extend its service to run the track circuitry with commercial power after the poles had been removed. However, it could not justify spending thousands of dollars to electrify low-power equipment because the expenditure would never be recouped. The Kansas City Southern rejected stand-alone diesels as a substitute power source because, as Taylor stated, "Diesel generators are an inefficient, outmoded technology." So in 1977 the railway chose to use photovoltaics to run its signals and switches whenever it could not connect to commercial power.[14]

Other railroads have followed suit, though it has taken time and effort to learn how to properly apply the new technology. In the northern states, railroads have struggled with learning how to work with photovoltaics. They need to keep the accompanying batteries well-charged during the short days and frequent cloudy periods of winter, while not overcharging them in summer when the sun rises at five in the morning and sets around nine at night. The Kansas City Southern solved this conundrum by hinging an extra panel to an existing one. The auxiliary panel can be easily raised and supported so it can work alongside the permanent panel on days of minimum sunshine; during summer, it can be dropped from the sun's view to prevent overcharging the batteries. When tracks run through a canyon, increasing the size of an array or elevating it compensates for the shading. "Over the years, the railroads learned a great deal about installing photovoltaic systems," James Le Vere testified, "and have made great strides toward increasing their reliability. Most railroads now consider the use of photovoltaics to power installations an attractive alternative to commercial power, depending upon utility charges to put in a line." The railroads' turn to solar did not arise "out of a sense of social responsibility," according to Le Vere. "It just meets their needs" better than any other power source.[15]

## Notes & Comments

1. Interview with Bernard McNelis. NASA's attempt to operate a terrestrial photovoltaics business serves as a cautionary tale. The scope of responsibilities required for success went far beyond the skills the agency brought to the task. The installation of a photovoltaic-powered refrigerator in Liberia illustrates why NASA failed. Jim Martz, a NASA engineer, remembers that he "took the thing out to the village, hooked things up, let the array charge up the batteries, plugged [it] in, and [it] started running. I tried to show the native technician what to do, but he didn't speak English and I couldn't talk to him. He knew nothing about batteries, he had no idea how a battery worked, how a refrigerator works, or anything like that. But we hooked the thing up, it ran, cooled down, and seemed all right. I did that in one day and went back to the capital, came back home, never heard another word about it, had no idea how long it ran, whether it's still running or not! . . . We never heard 'boo' after we'd installed them." Interview with Jim Martz, NASA–Lewis Research Center, Cleveland, OH.

2. Allan Rothwarf,  videotaped interview, 1994. (Courtesy Mark Fitzgerald.)

3. E. Berman, "Solar Photovoltaic Energy," *Hearings Before the Subcommittee on Energy of the Committee on Science and Astronautics*, U.S. House of Representatives, 93rd Congress, 2nd Session, 6 and 11 June 1974, Appendix, 88–89. Criticism of the government's role in the development of photovoltaics in no way implies that it has not played a positive role in helping the industry. For example, it provided testing devices which small start-up companies could not afford. The government's Jet Propulsion Laboratory's solar stimulator allowed photovoltaics companies to measure the power of their cells much more accurately than they could have on their own. "This was a very important early contribution," in Bill Yerkes' estimation. Furthermore, testing modules at JPL and publishing the outcomes protected consumers. Certification of a module by Jet Propulsion Laboratory took on the aura of the Good Housekeeping Seal of Approval for those purchasing photovoltaic products.

4. Interview with Bob Mitchell.

5. Interview with Arthur Rudin.

6. Interview with Bob Mitchell.

7. Ibid.

8. "Let the Sun Shine in," *Southern Railway Newsletter* (1979): 19.

9. Interview with Bob Mitchell.

10. Ibid.

11. John F. Norris, quoted in an ARCO Solar sales brochure, 1979?, "From East to West and North and South—Railroads Increasingly Use Solar Electricity." When Arthur Rudin left Solar Power for ARCO Solar, he brought much of the railroad business to his new employer.

12. Interview with James Le Vere.

13. Interview with Stanley Taylor.

14. Ibid.

15. Interview with James Le Vere. American railroads are not the only ones that run their safety equipment by photovoltaics. In the late 1970s, Australia, Italy, and Togo began to power railway signaling devices with photovoltaics as a way "to operate [them] with very reduced maintenance . . . offering a very valuable service at reduced costs." M. Trentini, "Photovoltaic Systems for the Railways in Italy," *Tenth E.C. Photovoltaic Solar Energy Conference* (Lisbon, Portugal, 8–12 April 1991) (Dordrecht: Kluwer Academic Publishers, 1991), 825.

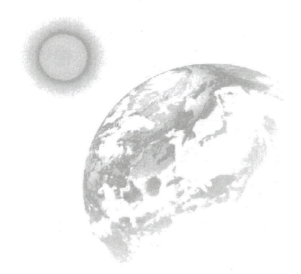

## Chapter Ten

# Long Distance for Everyone

When Bell Laboratories unveiled the silicon solar cell, Gordon Raisbeck, the director of its transistor program, predicted, "One field in which we can see immediate applications for it is communications—telephone, telegraph, radio, and television [where] the power need is small, but it is often needed in remote, inaccessible places where no power lines go."[1] Solar cells' first commercial success—facilitating communications between satellites and earth—soon fulfilled Raisbeck's prophecy. What could be more remote than space?

The U.S. Army Signal Corps, which pioneered the use of solar power in space, brought photovoltaics down to earth in June 1960 with the first coast-to-coast two-way radio broadcast generated solely by the sun's energy. A station set up in Ft. Monmouth, New Jersey, communicated to a similarly powered station at the El Monte, California, headquarters of Hoffman Electronics, the firm responsible for designing the solar arrays for the Vanguard and many subsequent satellites. Colonel Leon Rouge, the components research director of the Corps, came up with the solar-radio project to celebrate the one-hundredth anniversary of the organization. The *New York Times* quoted the colonel as viewing the experimental broadcast as the forerunner to making "radio and telephone communications practical in remote areas of Asia and Africa, where sun power is readily

*The head of the Signal Corps in Ft. Monmouth, New Jersey, made history in June 1960 by talking to engineers at Hoffman Electronics in El Monte, California, via a photovoltaic-run transmitter—the first of its kind.*

*The photovoltaic panels atop this Army truck served as the power source for the first solar-powered transcontinental radio broadcast.*

available, and where there is no economical means of obtaining electricity through power lines."[2]

Microwave networks did exist that could carry radio and telephone signals over long distances, with dish-shaped repeaters picking up and amplifying the signals. These networks usually consist of a series of parabolic dish repeaters mounted on large towers that act like relays. Each station, situated thirty or so miles (forty-eight kilometers) apart, has two dishes connected by cable—one to receive and the other to beam radio, telephone, and television signals, which are bundled as microwaves in a line-of-sight path to the next station. During the process of passing from one dish to the other, the microwave signal had to be shifted to a different frequency to prevent scrambling the outgoing and incoming transmissions. To modulate or change frequencies required a bevy of complex and temperature-sensitive electronic circuits that demanded much fine-tuning. A relatively large power unit, huge batteries or a generator, had to be brought in to run all this gear as well as the air conditioning and heating of a hut to house the electronics at suitable operating temperatures.

John Oades, a microwave systems engineer at GTE Lenkurt, a former subsidiary of the telecommunications giant GTE, was keenly aware of the shortcomings of dish repeaters. He had worked on quite a few in the 1960s. Since "many were up on mountaintops in the middle of nowhere" and their gadgetry "was complex enough to get in trouble anytime," Oades recalled, "you needed some way of accessing them, by cable cars or someone had to put in a road. Building that road could be pretty expensive and keeping it clear of snow in the wintertime would be very costly, too. On top of that, there was the need for power. When it was just not possible to bring in a power line to do that, inaccessible microwave sites had to have fuel or battery replacements brought in by helicopter."[3]

On a ski trip in the early 1970s, a solution to these seemingly irresolvable problems incubated in Oades' mind. Riding the chairlift always put him in a contemplative mood, and he started thinking about an episode of the old TV series *Dragnet,* the one where some kids had locked themselves in a refrigerator. When he began to consider possible uses for junked refrigerators, he realized that because they are insulated, sealed against moisture and radiation, "They would make great containers for microwave repeater equipment."[4] This insight led him to consider the kind of electronics needed, and then his mind really began to work. Suddenly he came to a startling revelation: Perhaps all that complex gear wasn't necessary at all.

Out of that ski-lift meditation came a revolution in microwave technology. Oades designed and patented a repeater that merely picks up a signal, amplifies it, and sends it on its way without the fuss and bother of changing or modulating frequencies. He discovered that using filters and covering the parabolic dish with a metal shield does away with any interference between incoming and outgoing signals, and this allowed him to eliminate most of the electronics. Bill Hampton, Oades' boss at the time, described the difference between Oades' invention and the old technology as going "from a rack of equipment to something you could put in your pocket."[5] Or in the words of the inventor, "This was the most minimalist way. . . . I simplified the technology about as far as you could."[6] What gadgetry remains, Oades culled from transistor technology developed for satellite-to-earth communications. The equipment to run Oades' repeater fits into a small cabinet about the size of a fuse box; it can work at any temperature encountered on earth; and it does not take much power or tending.

Because the system can operate for years without maintenance and requires only a few watts of electricity, Oades wanted a compatible source of energy, very reliable and compact. Therefore, choosing solar cells "came right at the beginning. In fact, if it weren't for photovoltaics," Oades swore, "I probably wouldn't have built the repeater."[7]

Hampton was very receptive to Oades' ideas for revolutionizing microwave technology, suggesting that they "do some testing, to see how well it works, and try to get the big boys [GTE executives] to finance it."[8] True to his word, in 1974 Hampton took some money from his engineering budget to build and test a prototype. In their spare time, the two set up a mini-microwave telephone network using Oades' repeater design. Serendipity simplified things. Hampton's house sat on a ridge overlooking the plant where the men worked. They put the prototype repeater into an unused microwave tower that was on top of the company's roof and relayed phone calls to and from Hampton's house and his hangar at a nearby airport.

Once the system was in operation, Hampton showed it to his boss and the corporate head of manufacturing for GTE. They witnessed this repeater doing the same job as a conventional one, but on no more power than that used by a nightlight. Interestingly, there were no generators or power lines in sight. When asked what gave the repeater its electricity, Hampton pointed to two Solar Power Corporation arrays placed above the dishes. The repeater's low energy consumption—1/100th of that used by the old technology—gave it "a unique advantage," Hampton told them. "It can be oper-

*John Oades (left) and Bill Hampton (right) check out the solar panels of Oades' solar-powered microwave repeater.*

ated entirely by solar power . . . permitting the repeater to be self-contained and self-powered [and for this reason] to operate in locations where a conventional repeater would be either completely impossible or inordinately expensive."[9]

The GTE officials more than liked what they saw. "They were overwhelmed," Hampton recalled. It didn't take long before GTE unveiled "the solar-powered, self-contained, super-economical . . . repeater" to the telecommunications industry in full-page advertisements, calling it "the most significant breakthrough in microwave transmission in the last thirty years." The claim raised eyebrows throughout the industry. GTE was suggesting that Oades' device compared in importance to the discovery of radar. As a consequence of the new repeater, GTE boldly announced, "The day when communications engineers had to think in terms of big towers, large power requirements, air conditioning, access roads, and all the attendant construction and maintenance difficulties for [repeaters] is drawing to a close. The first units are ready for shipment now."[10]

Navajo Communications Corporation bought the first unit. The company wanted to connect the community of Mexican Hat, Utah, with the rest of the world by telephone. The rugged, almost inaccessible terrain of deep canyons and sheer cliffs that surround the town had stood in the way of the long-distance telephone service that most North Americans in the 1970s took for granted. As a GTE executive observed, "It was one of those impossible situations of having to plow in miles of cable up and down the mountains and through a valley to a very small community. The investment on such a project would probably never have been returned. . . . Mobile

# GTE Lenkurt announces the most significant breakthrough in microwave transmission in the last thirty years.

## The solar-powered, self-contained, super-economical 2-GHz 700F1 RF Repeater.

GTE Lenkurt does it again. Introducing the 700F1 RF Repeater. So inexpensive, it brings the cost of line-of-sight transmission down to earth. So simple, it's installed in even the most rugged or inaccessible location, usually in a matter of days. So unique, it eliminates the conventional modulate/demodulate type repeater. So unconventional, it uses the same frequency, in and out, conserving the spectrum for expanded use. So advanced, it's in the process of being patented.

Passive billboard repeaters have solved the power and access problem in remote locations, but they present some formidable construction and cost barriers of their own. There is now another alternative—the 700F1.

The 700F1 is FCC type-accepted for repeating 36 message channels, plus 3 additional channels for order wire, alarm and control, in the 2-GHz frequency band. It is licensed to operate with the 70F1 Microwave Radio, typically utilizing 36A or 46A3 multiplex and 11A signaling systems. It knocks the props out from under the costs normally associated with conventional repeaters.

All active elements are in a weatherproof cabinet, rated for −40° to +140°F. No equipment hut. No air conditioning. No power generators. And the support structure can be wood pole, existing tower/building, or, in the worst case, relatively inexpensive guyed tower.

Where you can't reach the site by vehicle, a helicopter can easily get the complete assembly into place.

The 700F1 can be powered completely by solar cells. Less than 4 watts of power required. Storage batteries assure continued operation at night and in foul weather, and the batteries themselves are recharged by the solar cell on the next sunny day. If power is available at the installation site, you can save even more by eliminating the solar cells.

The entire system is 100% actively redundant with over 9 years MTBF per amplifier, utilizing advanced components whose reliability has been proven in the nation's space program. No field tuning—no levels to set.

Because of the 700F1, the day when communications engineers had to think in terms of big towers, large power requirements, air conditioning, access roads, and all the attendant construction and maintenance difficulties for narrowband radio systems is drawing to a close.

The first units are ready for shipment now—

**Get the full story.**
For full particulars, call (415) 591-8461, or write to GTE Lenkurt, Dept. C134, 1105 County Road, San Carlos, California 94070.

**Straightforward simplicity.**
Compare the complexities, construction and costs of a typical 2-GHz repeater. (see diagrams below)

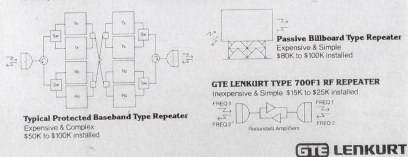

**Typical Protected Baseband Type Repeater**
Expensive & Complex
$50K to $100K installed

**Passive Billboard Type Repeater**
Expensive & Simple
$80K to $100K installed

**GTE LENKURT TYPE 700F1 RF REPEATER**
Inexpensive & Simple  $15K to $25K installed

FREQ 1    FREQ 1
FREQ 2   Redundant Amplifiers   FREQ 2

**GTE LENKURT**

*A GTE announcement, printed in trade journals, presenting their solar-powered microwave repeater to the telecommunications community. The photograph shows John Oades, its developer, tuning the broadcasting equipment, which fits into the small cabinet next to him. At Oades' right are the photovoltaic cells. The figures at the bottom compare the three microwave technologies available at the time.*

telephones would not have worked either. The high walls of nearby canyons would have effectively blocked out most signals."[11]

So Oades' repeater was installed on Hunts Mesa overlooking picturesque Monument Valley. The location gave the repeater a clear line of sight from the microwave terminal at Mexican Hat to a microwave link at Kayenta, Arizona, which tapped into the national telephone lines. The repeater and its installation cost Navajo Communications a fraction of what the alternatives would have. "It took less than two days' time to mount the pole structure, install the equipment, and get it on the air," reported an astounded J. Shepherd, CEO of Navajo Communications.[12] In the fall of 1976, Hunts Mesa became the first solar-powered microwave repeater site in North America and one of the first in the world.

*A view of the Hunts Mesa solar-run repeater, with a panorama of the rugged landscape that made its placement a necessity to provide Mexican Hat, Utah, with long-distance phone service.*

The repeater did not disappoint the people of Mexican Hat. Maintenance, as promised, proved to be minimal. According to Shepherd, this spared "field personnel from torturous travel to the site," a seventeen mile (twenty-eight kilometer) trip that takes nearly three hours by four-wheel drive under the best of conditions.[13] Fortunately, Oades and Hampton had worked through all the potential stumbling blocks at the home office. Through their experiments they had learned, for example, that you could boil a battery with a solar panel. "We ended up having to design regulators that would control overcharging," Hampton reported. Only one flaw marred the performance of the solar cells at Hunts Mesa: A solder joint cracked. Its repair took minutes. "Other than that," said Hampton, "the original solar modules are still up and running!"[14]

Oades regarded the Hunts Mesa location as an ideal showcase for the repeater. "It's . . . real isolated," he explained, "and it's a very good area for sun."[15] In the estimation of Navajo Communications CEO Shepherd, the Hunts Mesa installation "represents a major breakthrough in the economics of serving small population centers with microwave communications."[16]

Its success also boosted confidence in photovoltaics as a power source for repeaters and other isolated stand-alone applications. "The problem," Oades pointed out, "was that people were concerned about long-term reliability. Repeaters have to be very reliable. They're out there in the middle of nowhere with ice and snow, just terrible conditions. People didn't know how solar cells would stand up to all this. After a year of successful operation at Hunts Mesa, they had something to point to."

"It didn't take very long before the word got out" about the repeater and its power source, Oades added. "There are a lot of small isolated towns and villages . . . in the West that are hemmed in with mountains, where it would be too expensive to put in a conventional microwave system or telephone lines."[17] But thanks to the solar-powered repeater, they now could enjoy the same level of telephone service as urban centers.

Cuprum, Idaho, was another town that benefited from Oades' innovation. The inventor saw firsthand the positive impact his repeater had on the citizens of this small mountain enclave. "I'd be driving down the street with the owner of the [local] telephone company," Oades recalled, "and people would come out and stop him and thank him for providing them with long-distance service."[18] Prior to the installation of the solar-powered repeater, the hundred or so souls of Cuprum and neighboring Bear had to endure a bumpy fifty-five minute drive just to make a long-distance call.[19]

An improved solar repeater that could manage more than twelve hundred calls simultaneously was placed above Palm Springs on Snow Mountain, connecting the desert community to Ontario, California, where GTE's long-distance circuits for Palm Springs terminated. Unlike Hunts Mesa or Smith Mountain, where Cuprum's repeater stood, commercial power lines do run close by the Palm Springs' installation.

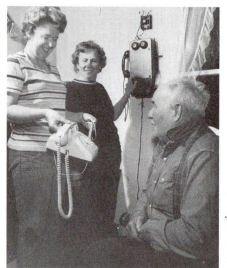

*Oades' solar-powered microwave repeater made it possible for rural Idaho residents Clarence and Beth Warner to call long distance. Joanne Wiggins, wife of the owner of the local telephone company, presented the Warners with their new telephone.*

But, as Hampton explained, "They weren't as dependable as one would like. Especially during winter when there'd be a lot of wind, rain, ice, and what have you. Up on Snow Mountain, the sun's a lot more reliable than Southern California Edison!"[20]

After six years on the market, GTE sold more than one thousand solar-powered repeaters, resulting in a multimillion-dollar worldwide business. "The big appeal and fanfare," according to Bill Hampton, "were the solar panels."[21] Since Oades' work intrigued those in the telecommunications business, both Oades and Hampton received invitations to conferences and meetings throughout the world to discuss their work. One excursion took Oades to Australia, where professionals from "Down Under" had good reason for their interest: Australia is roughly the size of the United States, yet in 1970 fewer than twelve million people lived there. Its highly developed standard of living called for modern telecommunications links, but the networks the Australians wished to build had to traverse vast, sparsely inhabited—albeit sunny—terrain. To accomplish this feat, the Australian government generously funded Telecom Australia, the quasi-public agency in charge of the nation's communications services. The national policy established early in the 1970s was to try to provide every citizen, no matter how remotely situated, with telephone and television service comparable to that enjoyed in the larger population centers.

Michael Mack, a power engineer for Telecom Australia, explained the dilemma his agency had to resolve. "In trying to get telecommunications to outlying properties, and we had a lot of them that are really quite remote, we found that one of the biggest problems was a reliable power supply." Telecom engineers therefore brought in thermoelectric generators, diesels, and wind-driven machines; sometimes they connected their equipment to the customer's own power source. But none of this panned out as well as Telecom Australia had wished. "They weren't reliable," Mack complained. "The problem was you were dependent on somebody actually maintaining the wind generator or keeping gas bottles replenished. Or if it was the customer's power supply, the quality of that power was very uncertain."[22]

On the lookout for something better, Arnold Holderness, a senior power engineer for Telecom Australia, kept his eyes open while on trips to America, Europe, and Japan. At the headquarters of Sharp, the Japanese electronics equipment manufacturer, Holderness saw some literature about the photovoltaic modules that the company had just started to produce. He ended up taking some Sharp panels home. "We were alive to the need for remote power supplies, and with this government mandate to hook everyone up, we had to think innovatively," Holderness said.[23]

The high price of the Sharp modules, about $100 per watt, severely limited their application. But a year or so later, the appearance of Solar Power modules, which were five times cheaper than the Japanese product, gave Telecom Australia the green light to put photovoltaics to work. The assistant manager of the agency told the Australian Senate Committee on National Resources in 1976, "At $20 per peak watt [the price of the Solar Power modules at the time], photovoltaic conversion is an attractive proposition for small power supplies in remote areas [and we have] about 20 small solar systems installed." Each system consisted of a telephone, a transmitter, and a receiver powered by low-wattage Solar Power modules. Calls were beamed to and from a telephone exchange that was connected to the national network. The solar-powered phones worked so well that after several years of service the staff at Telecom Australia could state with great confidence, "Using direct photovoltaic conversion has become a viable and preferred power source for [remote telephones]."[24]

By demonstrating through these micro-installations "a cost benefit over any other primary power source suitable for the job," the initial solar program of Telecom Australia was deemed to have been successfully completed by the summer of 1976. Telecom engineers had gained the "confidence and experience with [the] new power source" they needed. They were now ready to "design and develop large solar power systems" that would link distant towns with the rest of Australia's phone and television networks.[25]

The dramatic drop in the amount of electricity needed to run large repeaters also helped Telecom Australia to consider powering entire telecommunications networks by the sun. A large dish repeater that previously had required a kilowatt to operate now took only one or two hundred watts. Improved solid-state equipment brought about the breakthrough, as did housing the circuitry equipment underground where temperatures remained mild year round, thus eliminating the need for air conditioning or heating.[26] These repeaters, though, consumed much more power than did Oades' invention because the Australians needed to relay television and telephone signals over hundreds of miles, something the Lenkurt design could not do. By 1976, the power necessary to run a large-scale repeater—one that could carry hundreds, if not thousands, of calls at a time—fit amazingly well with the loads that photovoltaics could economically handle. As Telecom Australia reported that year, "[T]o supply 100 watts in a remote area, without mains [utility] power, solar means are cheaper than any other source."[27] And just a year later, an issue of *The New Scientist* announced that photovoltaics "can be justified economically for loads up to 200 watts" in faraway places.[28]

The significantly reduced maintenance requirements of the new microwave dishes also expanded Telecom's interest in solar electricity. A typical solid-state-equipped microwave repeater—that is, a repeater using electronic components that need neither heat nor moving parts to operate—built during the mid- to late 1970s could run pretty much trouble free for ten years. The same could not be said for the repeater's usual power source when no commercial electricity was close at hand. Diesel generators demand major servicing on a yearly basis and many refueling visits.

*Above: Telecom Australia's first remotely situated photovoltaic-powered telephone system, installed in 1974. Its success led to Telecom Australia's extensive use of photovoltaics in the 1970s to power microwave repeater systems. Right: A close-up of the system's solar power pack.*

Considerable time and labor had to be expended on the diesels, while the dishes themselves required none at all. The engineers responsible for maintaining the microwave dishes were open to finding a self-sufficient power source. In other words, they were ready for photovoltaics.

Despite these compelling reasons to turn to solar, Telecom Australia's decision to rely on the sun as a major power source surprised a lot of people. "It was considered a fairly risky thing in the 1970s," Michael Mack recalled, "quite a novelty."[29] Success hinged on selecting the right solar equipment.

Holderness and Mack sought a photovoltaic device that could duplicate the solid-state equipment's stellar performance. They were looking for a module that would last at least ten years and that would require only minimal care, a service visit every six months, while operating under severe conditions—in deserts where daytime temperatures rise to more than 120°F (49°C) and drop to freezing at night or in the tropics where the humidity never falls below 90 percent. With this in mind, they conducted an exhaustive series of accelerated environmental and laboratory tests.

Preliminary field tests ruled out almost every module on the market because, by then, most used silicone as the encapsulant and cover. Telecom Australia had discovered that parrots and cockatoos native to Australia consider silicone a delicacy. Another problem with all-plastic modules was that dirt kicked up by wind storms readily accumulated on the silicone, shutting out sunlight from the cells. Glass provided the remedy: It would both protect the plastics inside from pesky bird beaks and let dirt slide off its slippery surface. Holderness and Mack therefore decided that any module Telecom Australia used would have to have a glass top.

In the mid-1970s, this requirement allowed them only one choice—the RTC module built by Philips Electronics. But that was acceptable because Philips had a factory in Australia and the company's engineers were thus able to work closely with Holderness and Mack. The first RTC module that underwent the tortuous testing procedure that Holderness had developed—which included temperature cycling from -13°F to 185°F (-25°C to 85°C) and steambaths—completely fell apart. In correcting the problems that had caused the module to fail, Philips developed a corrosion-resistant cell. Its engineers also reduced the number of materials placed inside the panel. This checked the breakage of cells and cell connections that had occurred with the various packaging materials expanding and contracting at different rates due to the dramatic rise and fall in temperature. The refined RTC design became

*One of Telecom Australia's many solar-powered microwave repeaters, which provide rural Australians with first-class telephone and television services.*

the standard power pack for Telecom Australia's remote installations.

With the modules, batteries, and control equipment having passed muster, Telecom Australia felt ready to construct its first large-scale solar-powered telecommunications system. Thirteen solar-powered repeaters went up in 1978, each twenty-five miles (forty kilometers) apart. They connected the existing diesel-powered network at Tenant Creek to the resort town of Alice Springs, linking in the process such colorfully named intermediate points as Devil's Marbles, Tea Tree, and Bullocky Bone to Australia's national telephone and television service. People in these and neighboring towns, such as 16 Mile Creek and Warraby, no longer had to dial the operator for long distance and shout into the phone to be heard. Nor did they have to wait for news tapes to be flown to their local stations, viewing them hours after the rest of Australia had.

Telecom Australia's high standards paid off. The Tenant Creek–Alice Springs solar-powered system worked well. "The concept proved so successful," Michael Mack and colleague George Lee reported ten years later, "that Telecom Australia went on to install seventy similar solar power packages throughout its network."[30] All the installations had "gratifying results" as "there had been no system failures" after ten years of operation.[31] The Kimberley Project, which links distant, but booming, northwestern Australian towns, remains the world's largest sun-driven microwave system. Forty-three solar-powered repeaters, spaced about thirty-five miles (fifty-seven kilometers) apart, span fifteen hundred miles (2,420 kilometers).[32] Not only do the repeaters bring long-distance telephone service to the towns along

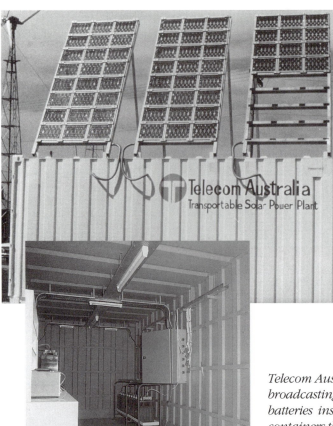

*The success of its solar-run telephone systems led Telecom Australia to construct large-sized photovoltaic power plants to run microwave repeaters.*

*Telecom Australia placed the broadcasting equipment and batteries inside the shipping containers that supported the photovoltaic panels*

their line, but people at sheep stations and isolated homesteads and communities within thirty miles (forty-eight kilometers) of their range can connect to them via solar-powered phone systems.[33]

Telecom Australia's large-scale solar projects made it one of the world's major purchasers of photovoltaics from the late 1970s through the 1980s. A photovoltaics firm doing business with the telecommunications agency could count on doing well if its product could measure up to Philips' benchmark module. Solar Power Corporation reached that plateau, incorporating a glass top and other robust features into its modules. These improvements could not have come at a better time for the Australians. Their demand for very dependable modules had grown exponentially, and having another suitable supplier kept the market competitive. Other manufacturers began to produce extremely durable modules, too, allowing Michael Mack to inform an international telecommunications conference in 1984, "The [photovoltaics] industry has matured in recent years, with numerous high reliability modules now in the market."[34] The increasing number of reliable photovoltaics products to choose from made it easy for Telecom Australia to continue using solar electricity after Exxon and Philips dropped out of the business. The Australian subsidiaries of two American firms, Solarex and Tideland Signal, kept the Telecom group well-supplied through the 1980s.

At the same conference, Mack also told his colleagues, "We have advanced from the stage where solar power was considered an exotic source to where it now plays an important role alongside conventional power supplies."[35] The work of Telecom Australia helped the photovoltaics industry reach this high stage of acceptability. As Mack put it, "The Australian experience gave people all over the world [the] confidence to use photovoltaics. Telecommunications groups could confidently point to Australia and say, 'This is what Australia is doing and look at the harsh environment they succeeded in.'"[36] Or in the words of Arnold Holderness, "We were

showing the world how solar power could be used in a big way out in the field."[37] Indeed, the world caught on. One telecommunications expert told colleagues in 1985 that photovoltaic systems have "become the power system of choice [for] remote communications."[38]

*Photovoltaics allows this isolated Australian homestead to communicate with the outside world as easily as urban dwellers.*

## Notes & Comments

1. G. Raisbeck, "The Solar Battery," *Scientific American* 193 (December 1955): 110.
2. "Solar Radio Makes Coast-to-Coast Hook Up," *New York Times* (24 June 1960): 7, column 2.
3. Interview with John Oades.
4. Ibid.
5. Interview with Bill Hampton.
6. Interview with John Oades.
7. Ibid.
8. Interview with Bill Hampton.
9. W. Hampton, "A New Look at RF Repeaters," *Telecommunications* 10, no. 5 (May 1976): reprint.
10. "GTE Lenkurt announces the most significant breakthrough in microwave transmission in the last thirty years." Advertisement.
11. Interview with William Hunter.
12. J. Shepherd, "Navajoland's Solar Powered RF Repeater is First in the Nation," *Telephone & Engineer Management* (15 November 1976): reprint.
13. Ibid.
14. Interview with Bill Hampton.
15. Interview with John Oades.
16. Shepherd, "Navajoland."
17. Interview with John Oades.
18. Ibid.
19. K. Wiggins, "The World at Their Fingertips," *Telephony* (16 April 1979): reprint.
20. Interview with Bill Hampton.
21. Ibid.
22. Interview with Michael Mack.
23. Interview with Arnold Holderness.
24. Standing Committee on National Resources, Australian Parliament. "Reference, Solar Energy: Official Hansard Report/Standing Committee on National Resources" (1976): 2.1706, 2.1708. 2.1730.
25. Ibid., 2.1706, 2.1730.
26. M. Mack, "Solar Power for Telecommunications: The Last Decade," *International Telecommunications Energy Conference* (New Orleans) (New York: IEEE, 1984), 274; M. Mack, "Solar Power for Communications," *Telecommunications Journal of Australia* 29, no. 1 (1979): 390.
27. Standing Committee on National Resources, Australian Parliament, "Hansard Report," 2.1730.
28. "Australians Dial Calls Courtesy of the Sun," *New Scientist* (9 June 1977): 590.

29. Interview with Michael Mack.

30. M. Mack and G. Lee, "Telecom Australia's Experience with Photovoltaic Systems in the Australian Outback," *Telecommunication Journal* 56 (1989): 514.

31. D. Eskenazi, "Evaluation of International Photovoltaic Projects," Sand 85-7018/ 2 DE 87 002943, B-10, 1986.

32. Mack and Lee, "Telecom Australia's Experience," 514–15.

33. M. Mack, "Solar Power for Communications," 279.

34. Ibid.

35. Ibid., 274.

36. Interview with Michael Mack.

37. Interview with Arnold Holderness. Bill Yerkes agrees with Holderness and Mack's assessment of Telecom Australia's major contribution to the progress of photovoltaics. Yerkes credited Telecom Australia as the real "leader in applying" the technology. In an internal document published in the 1980s, Solarex made a similar assessment of Telecom Australia's pivotal role, citing the Australians as primary to "the development of solar power as a practical energy source." L. Permenter, "In Touch in the Outback," *Sunworld* 12, no. 3 (1980): 80.

38. I. Garner, "Design Optimisation of Photovoltaic Telecommunication Applications," *The Conference Record of the Eighteenth IEEE Photovoltaic Specialists Conference—1985* (Las Vegas) (New York: IEEE, 1985), 275.

## Chapter Eleven
# Father Verspieren Preaches
## the Solar Gospel

The twentieth century had never witnessed drought and famine of such immense proportions as that which hit the African states situated directly below the Sahara in the 1970s. "People and livestock fell like flies" as a consequence.[1] The little-known countries of Chad, Mali, Mauritania, Niger, and Senegal gained global recognition as the world learned of their plight.

One European, Father Bernard Verspieren, could not just read about the tragedy, shake his head in pity, and continue his daily routine. As a Catholic priest, Verspieren ran a mission in rural Mali. Every day women, who had walked tens of kilometers with empty containers balanced on their heads, passed by his church in search of water. He saw robust people and livestock decline into skeletons. He saw the villages in his parish empty as people fled to survive. "It became increasingly uncomfortable for me to say my breviary," the priest confessed, while watching neighbors and neighborhoods slowly die.[2]

His opportunity to help came when Malian authorities asked Verspieren to form a private company to drill wells in the region of San where he lived, realizing that the survival of the country depended on pumping water. Verspieren's success in establishing and running two local agricultural schools had impressed those governing the African state. They saw him as someone who could get things done.

Verspieren formed a nongovernmental organization, which he named "Mali Aqua Viva"—"Living Water for Mali"—to tap the great rivers that flowed deep beneath the dry sands. The government would provide 80 percent of Mali Aqua Viva's funding; Father Verspieren would have to come up with the rest. Though his order, the White Fathers of Africa, demands a vow of poverty, it allows members to spend their own money on projects they wish to pursue. Fortunately, Verspieren came from a very wealthy family and he generously endowed Mali Aqua Viva. He also tapped his extensive European contacts for additional financing.

Through Father Verspieren's efforts, two drilling teams arrived in Mali in 1975. In the following five years, 620 wells were drilled and only 120 came up dry. But what dogged the operation from its beginning was the lack of a reliable power source to run the pumps. Regular commercial power was not available in the villages where most of the wells had been drilled. And even if it were, it was notoriously unreliable. In the capital of Bamako outages occurred all the time.

Diesel generators were not a good choice either. The hospital at San had purchased one and, like so many other diesels in Africa, "It was usually broken down or in want of fuel." As a consequence, "The hospital had neither electricity or running water." Sporadic supplies of fuel, too few mechanics, and little in the way of spare parts caused San, and most of Africa, to become, as Father Verspieren wryly put it, "a burial ground of motors."[3]

In desperation, having found nothing better, Father Verspieren settled on human power. Hand and foot pumps require a lot of work—twelve hours a day, every day of the year—for meager results: The water just trickles out. For a village of one thousand, manually powered pumps would provide barely enough water for drinking and none for crops or livestock.[4] Worse, these pumps frequently break down. "Practically speaking there is no such thing as a long-lasting hand or foot pump," Verspieren complained, after using them for several years. "Their maintenance is very great when compared with the amount of water they extract. I have two maintenance teams who have to constantly crisscross our pumping region without stopping to keep them going."[5]

Despite the problems associated with manual pumps, many international aid "experts" in Europe and America were smitten by them and considered them the answer to Africa's water needs. Verspieren thought it was crazy to expend so much effort for so little water. He dismissed the supporters of this method as neocolonialists who were really thinking, "If Africans don't work more, then the whole continent is headed toward disaster."[6] With so much water underground and with so much sun overhead, many in Mali, and the rest of West Africa, began to wonder, as did one United Nations' engineer, "Why solar energy should not be used to pump water to save thousands of people and cattle from dying of drought."[7] In the late 1960s and early 1970s, engineers had considered using the sun to power water pumps, but they unanimously had dismissed photovoltaics as a possible power source. "At first glance, [photovoltaics] appears very attractive," wrote P. Masson, a French engineer working in West Africa.

*A Malian child operates a hand pump.*

But, considering the price, he decided that the "direct conversion of solar energy into electrical energy is, in the immediate future, limited to the space industry."[8] Instead, he and almost every one in the renewable energy field shared the prevailing view that "for the African native . . . [solar] power generators with moving parts appear to be a more practical solution to [the] problem."[9] The international aid organizations and the French government listened to the experts and embraced solar thermal pumps, spending hundreds of millions of francs to construct them. Although almost identical to the first sun motor Frank Shuman had built in 1907, people regarded the solar pumps as novel. As with Shuman's first sun engine, a fluid with a low boiling point passed through glass-covered panels, collecting enough heat to vaporize it. The resulting steam would run an engine that would activate a pump.

Verspieren, always on the lookout for a better way to run his pumps, visited one of these solar thermal pump installations in Mali. He immediately saw that it would never work in Africa. Each thermal pump required a resident engineer on duty twenty-four hours a day to make adjustments and repairs. In Africa, such skilled personnel would not be available. To Verspieren, the "new" solar pump was just like a diesel pump. Because it had an engine, "You fall into that old abyss of motors, of turbines," he

*The complicated moving parts shown in this photograph proved to be a fatal flaw to solar-heat powered engines in Africa.*

pointed out.[10] Its extremely inefficient conversion of solar energy into useful work also mandated a bevy of solar collectors that weighed many tons and occupied more than a thousand square yards. Verspieren wondered how such a bulky object could be carried to villages where roads are always questionable at best. Or worse, what if the sun pump were in place and the well went dry? Imagine, he thought, trying to take that monstrosity apart to reassemble somewhere else! What the Africans got, in the words of one critic, "was [an] extremely cumbersome, material-intensive [technology] . . . whose costs were high and reliability in remote locations very doubtful."[11] Terry Hart, the former chief technical advisor to the Malian National Solar Laboratory, called solar thermal pumps "a classic example of European laboratory technology which is not at all adapted to field conditions."[12] Verspieren would have nothing to do with them. And that was smart as none of them ever worked for very long.

Despite the inherent shortcomings of the solar thermal pumps, throughout the 1970s they stayed quite the rage among renewable energy technologists as the hope for the developing world. A few heretics, however, shared Verspieren's disdain. Among the dissenters was Dominique Campana, who graduated from engineering school in France in the mid-1970s. She contributed to the development of the world's first photovoltaic-powered water pump as part of her doctoral thesis.

While a university student, Campana had developed a great concern for the environment. "It was the 1970s," she explained. "Young people like myself were interested in all that was natural, protecting nature, and relying on natural sources of energy."[13] To determine if professional opportunities existed that were compatible with her ecological philosophy, she attended every conference on renewable energy technologies that she could. It was a stroke of luck that UNESCO held its Solar Summit in Paris in 1973, where the leading experts in the field converged to map out the world's solar future. She went, of course, and at the end began to seriously consider a career in solar energy.

Her interest in environmental matters also led her to examine the difficulty that people, especially those in out-of-the-way places, might have in obtaining enough water. Water and solar energy, she noted, shared a singular irony. Although regarded as gifts from God, elaborate and expensive mechanisms often must be devised to make them useful or available.

Campana learned how to best use solar energy to solve future water shortages during a lecture given by Dr. Wolfgang Palz and in informal talks with him afterward. Known in France as "Mr. Solar" because of his great knowledge and strong advocacy of solar energy in general and of photovoltaics in particular, Palz had opposed solar thermal pumps from the beginning. "The technology was wrong for physical reasons," he steadfastly maintained. "The efficiency of such a thing is terrible . . . far below 1 percent."[14] The logic of his arguments and the charisma of his delivery led Campana to photovoltaics. She took the challenge of applying what hitherto had been primarily confined to outer space to one of the most humble purposes on earth—the provision of water. Photovoltaic pumps, she was convinced, would have distinct advantages over solar thermal ones: "1. Complete autonomy; 2. Simplicity of construction; 3. Higher efficiency and lower solar collector area; and, 4. Lower weight and simplicity of transport."[15]

*Dominique Campana, designer of the world's first commercial photovoltaic-powered water pump.*

She enlisted the support of Pompes Guinard, France's leading pump manufacturer; Philips donated the photovoltaic modules. With Guinard engineers, a prototype pump was built that would work with photovoltaic panels in a trying environment and that would operate from the direct current which the cells produced. The efficiency of the pump was improved, which reduced the number of panels necessary to drive it, which kept costs down. To minimize maintenance, they made the components sturdier. Both Pompes Guinard and Philips saw this work as an opportunity to develop a future market for their products.

It was decided to install the pump in the mountains on the island of Corsica, where Dominique Campana lived. She could then watch the system and make necessary adjustments. In addition, Corsica's rugged isolated terrain came the closest in Europe to replicating the conditions in Africa, where, she hoped, the pumps would one day operate. A former Parisian who had made his career in computers, but who had decided to drop out and take up shepherding in Corsica's mountains, lent his property for the installation. He needed water for his sheep.

The pump worked flawlessly, Campana wrote, providing water for "300 ewes and an agricultural operation." She also saw that the promise of her pioneering work went far beyond the Corsican installation: "[I]n the

*Solar scientists from all over the globe came to Corsica to visit the world's first photovoltaic-powered water pump.*

remote and arid areas where the water problems are of fundamental importance, such a system finds interesting applications."[16]

Word of the revolutionary pump spread throughout the world solar
community. Those seriously interested in solar water pumping climbed the
Corsican mountains to see the apparatus at work, though the hilliness of
the site and its out-of-the-way location did not make the trip easy. Father
Verspieren was among those who made the pilgrimage, and he found it
well worth the effort. Everything about the pump impressed him. "Seeing
an electrical current produced without moving parts, without fuel, without
a generating plant, nor [requiring the constant attention of] a technician
was convincing. For me," the priest added, "it was love at first sight."[17]

The pump's output also caught Verspieren's imagination. Under less
than optimal conditions, at a latitude shared by Chicago, it drew twice as
much water from a much deeper well than did any of his manual pumps.[18]
By contrast, the region of Mali where Verspieren worked lies far below the
Tropic of Cancer, less than fifteen degrees from the Equator. A photovoltaic pump under the Mali sun would surely trigger the regeneration of the
land, Verspieren believed, leading him to declare the development of the
photovoltaic pump a miracle.

Verspieren returned to Mali very much inspired, "hypnotized by the
cell."[19] However, the day-to-day operation of the Mali Aqua Viva network
of manual pumps kept him too busy to immediately apply the conversion
he had experienced at the sheep farm in Corsica. For the moment, he was
convinced, his full attention to the current pumping system would at least
keep his diocese alive. Or so he believed, until the afternoon when his
faith was shattered by the deaths of two Malian women. On their way to
market, they had expected to find enough water at one of Mali Aqua Viva's
roadside hand pumps to keep them going until they got to their destina-
tion. Unbeknownst to them, the pump no longer worked. Because of the
breakdown, the women died of thirst somewhere along the road.[20]

Their deaths roused Verspieren to do what he knew had to be done.
There surely would be more casualties if they continued to rely on manual
pumps. The thirst of his people sent him back to Europe, this time to beg,
cajole, coax, wheedle—whatever it took—to get the funds to purchase
photovoltaic pumps for Mali. Europeans listened to the solar priest. He
went straight to their hearts and then to their pocketbooks. As an American
solar engineer witnessed, "Verspieren grabbed anyone's arm that he could

find to ask for money. 'If you can't give me a thousand dollars,' he'd say, 'give me a hundred. Why not?!?'"[21] His people needed water and Father Verspieren believed with all his heart and mind that only the sun that had made the drought could stop it.

The roving beggar, as Verspieren calls himself, harking back to the medieval precedent of the mendicants, monks who walked the streets for alms to support their monastery, had gained the support of several international charities. With their money, he bought the first set of panels from Solar Power Corporation and a pump from Pompes Guinard. He chose Nabasso, a large village near Mali Aqua Viva's base in San, for the first African photovoltaic water pump installation because its people had taken the initiative "to do things for [themselves], such as build [their] own school and dispensary."[22]

Early one hot morning, after the technicians had connected the last pipe and wire, the people of Nabasso were amazed. They heard water gushing up the well pipe, but without the noise of a generator. Nor did they detect even a hint of smoke. Yet water began to fill their too-long empty reservoir. The stunned crowd had expected another dry, dismal day brought on by the scorching sun. Who could believe that the flow of water would grow stronger as the sun climbed in the sky? For the villagers, "It was like magic," stated Guy Oliver, a French solar engineer who worked with Mali Aqua Viva back then. "At first, it was impossible to explain to them that the sun was running the pump. But when night came and the pump stopped, they started to understand."[23]

*Young Malian boys watch a photovoltaic-driven pump fill their village's formerly dry cistern with water.*

*Father Bernard Verspieren explains to visiting dignitaries how solar modules will rescue Mali from its terrible drought.*

The people of Nabasso sensed that history was being made in their village. Father Verspieren explained the significance of the new technology as he preached his gospel of the sun at the pump's official inauguration: "What joy, what hope we experience when we see that sun which once dried up our pools now replenishes them with water! Everywhere in the world the big problem with wells was to find reliable means of drawing water out of them. . . . This time you have the answer right before your very eyes. Solar power is the answer! It will be your salvation. . . . This extraordinary discovery is no longer a dream: You've seen it, touched it, listened to it—not in a laboratory, but in your own backyard!"[24]

Since that momentous morning, the village has changed. People celebrate, rather than dread, the sun as it begins its daily arc. Verspieren observed, "In the morning, all the women come to do the laundry; there are acres of clothes drying in the sun. The toilette of these ladies, their culinary preparations, are all done here. The cows no longer have to travel miles for water and are in good health, and the manure, which had been lost, helps fertilize the local crops. . . . And the children, who used to cast pebbles in the [reservoir] to determine whether or not there was any water left . . . now come [there to sail] their miniature dugout canoes."[25]

Nabasso became the symbol of hope in a drought-ravaged land. Not bad for a place that two years before had been slated to become a ghost town. Word spread throughout Mali and beyond about Nabasso and the white priest endowed with special powers. "He is something like a sorcerer," they'd say. "He draws water using the sun's rays."[26]

The Malian villagers became ready converts to photovoltaic water pumping, but outside Mali, especially in Europe, skepticism abounded. Westerners scoffed at solar cells as playthings, not capable of real work. Others waited for the experiment at Nabasso, like all "white man" projects, to fail one day or the next. But it didn't. The people of Nabasso, as Verspieren expected, took good care of their installation. They built an adobe wall around the modules to keep livestock from wandering in and damaging the system. They washed the panels whenever necessary.[27] For over three years the modules and pump at Nabasso enjoyed a perfect track record, something that could not be said for any other type of machinery brought from the West to Africa.[28]

Elsewhere in Mali, the water crisis worsened. The village of Woloni, like so many others, had to cart water in from streams miles away. Satisfying the thirst of twelve hundred people in this way could continue for only so long before the villagers would lose their valuable herds of cattle and leave. Such dire straits forced the village leaders to approach Verspieren for help. The priest told them that he would bring in his drilling trucks and, yes, they could get the solar pump they had requested. However, certain

*The "perfect marriage" of sun and water, made possible by solar cells, provides the villagers' livestock with an ample water supply.*

things had to be agreed upon first. They would have to pay Mali Aqua Viva $5,500. Villager participation in the drilling and construction of the reservoir was mandatory. The village must also feed and house the project team for as long as work continued. A system for managing water distribution had to be devised, including a method to systematically collect money from users to cover maintenance costs. Providing nontechnical upkeep of the equipment—keeping the modules and the site clean—was also a prerequisite. And the people were to keep their hands off the wiring and the pump. In return, Mali Aqua Viva promised to install a solar pump and keep it running.[29] If the villagers did not like the terms, so be it, Verspieren said with a shrug. There were many others who would be only too happy to comply, and he showed them a waiting list to prove his point.

Verspieren established these terms—and stuck by them—because he had lived long enough to know the significance of the Malian proverb, "Whatever you do, you do it for yourself." As far as the priest was concerned, only if the village invested its own money, sweat, and time would the people regard the installation as theirs. Then and only then would it become a valued possession and treated with attention and care.

The investment Woloni put into its solar pump returned bountiful dividends. It brought them the same prosperity that had blessed Nabasso. With more water available than ever before, adobe bricks could be made in greater numbers, allowing the enlargement of huts. Extra water made it possible to set up small gardens, whose produce sold well in neighboring markets. The cash bought clothes, transistor radios, bicycles, and other modern goods the villagers had yearned for. "Now the young people feel good here," a satisfied villager remarked, "they have no desire to leave."[30]

Mali Aqua Viva also installed two manual pumps alongside the solar pump. On the rare cloudy day, these pumps would supplement the sun's decreased output. Or if the solar pump broke down, they could turn to foot and hand power to maintain their water supply. And because the manual pumps were used only in emergencies, they remained in working order much longer than before.

By 1981, when Woloni got its solar pump, Mali Aqua Viva had completed twenty-five similar installations. The forty-five kilowatts of photovoltaic power used in Verspieren's diocese made Mali Aqua Viva, along with Telecom Australia, the largest commercial purchasers of solar cells in the world at the time. This was no small accomplishment for one of the world's poorest countries, especially considering that the villages participated in the purchase of pumps and the installation of auxiliary equipment. "Their involvement is very nearly one-quarter of the investment," Verspieren noted. "It is the first time in thirty-one years in Africa that I have seen that."[31]

"Of all of our means of pumping," the priest declared, after almost five years of experience with photovoltaics, "solar pumps are the most efficient [and] the most reliable."[32] Still, some installations required technical maintenance, although the majority of the problems were with the pumps rather than the modules. Because the wells were drilled by ramming, not boring, it left them angled rather than perfectly straight. Consequently, the drive shaft of the pump had to be bent to fit the contour of the well. This improvisation put stress on the bearings, which wore out more quickly than expected.

Fixing one of these long shafts was no picnic, as Momadov Diarra, a Malian engineer in charge of repairs in the early days, can vouch: "When you had a problem with the shaft, it was very painful. You're out in the desert, its over a hundred degrees, and it takes a whole day to pull it out with the help of four or five people."[33] A large derrick truck had to be called out to haul the shaft back to the shop, where specialized equipment would be used to make the necessary repairs. From 1980 to 1983, seventy-six such time-consuming service runs had to be made.[34]

Problems such as these would have spelled doom for Verspieren's solar project had he not built up a stockpile of spare parts, an ultramodern repair facility, and a highly trained staff of African and European technicians. Also, the priest kept the installations within a reasonable distance from the repair facility so that any breakdown could be readily attended to. Terry Hart praised the priest for such foresight, stating, "Verspieren was wise enough to install all his pumping systems in a cluster within about a one-hundred kilometer (sixty-two mile) radius of his headquarters, so that with one base he was able to service a whole region without having to run from one end of the country to the other, one of the shortfalls of many earlier developmental projects."[35] Verspieren knew too well that no mechanical system is exempt from trouble and that unless a broken pump is quickly repaired, an entire community can perish.

The appearance of a superior pump in the early 1980s also helped Mali Aqua Viva's solar operation. Grundfos, one of the world's largest producers of high-quality pumps, had been eyeing the solar field as one of great sales potential. Their pump had proven its worthiness in hundreds of thousands of locations that used commercial AC electricity. By developing a reliable inverter that transforms the DC current produced by the solar cells into AC, this same pump could work with photovoltaics.

The Grundfos pump was elegantly simple: It had no shaft. Upon learning this, the Malian technicians gave a collective sigh of relief. Wiring protected by flexible plastic connected the inverter, which was placed above the well, to the immersed motor and pump. The pipe that carried the water was also made of supple plastic. Thus, everything that went into the well naturally conformed to its contour—and no part of the immersed material could corrode. The motor and pump were stainless steel, and the motor needed little attention because it was self-lubricating.

Moved by the same reasoned impetuosity that had directed his decision to commit to photovoltaics, between 1983 and 1986 Verspieren replaced all Guinard pumps with Grundfos (and later, with Grundfos-like equipment manufactured by Total Energie). As he explained, "For me, cost is secondary. What is of primary importance is reliability because on that depends the viability of our people."[36]

Once again, Verspieren's eye for good technology pointed him in the right direction. Each new Grundfos pump required a service call once every two and a half years; the Guinard installations needed six to ten visits annually.[37] Of all the repair calls made on Grundfos immersed pump sets, 90 percent "don't call for a change in parts, but only require a simple cleaning of the pipes or a quick fix of the wiring," affirmed Jerome Billerey, an engineer formerly in charge of maintenance for Mali Aqua Viva's solar equipment.[38] Since the wiring and pipe do not weigh much, a four-wheel drive vehicle can easily pull them from the well in a matter of minutes, and they can be reinstalled in a jiffy. Even extensive repairs can then be done onsite.

Improvements in photovoltaic modules also added to the reliability of Mali Aqua Viva's solar program. Like the Guinard pumps, the first three modules worked but needed a lot of care. Only a coat of clear silicone protected the silicon solar cells and the silicone in which they were embedded. Due to early problems with an acrylic top, it was discarded: The top and the silicone inside expanded at different rates under intense sunlight, which caused the silicone to separate from the cover or cracks to appear at the edges of the panel. Lacking substantial protection, the hot sun would soften the exposed silicone. When winds whipped through the desert, they would blow sand into the softened material. Worse, as the desert cooled at night, the silicone would harden. The constant contraction and expansion eventually broke the connections between the cells. Also, moisture from the high humidity penetrated the modules and slowly corroded the wiring inside. Consequently, the panel's power gradually dropped.

Verspieren flew from Mali to Cannes, France, for an international meeting of photovoltaics specialists to tell them about such problems in hope that his testimony "would motivate" improvements. He told the group quite frankly that Mali Aqua Viva's honeymoon with photovoltaics was over: "We cannot shut our eyes to reality. We must inform you that [in Mali] the sun is harsh and the cells are fragile. We are ready to pay dearly for modules," he entreated the conferees, "but please, don't try to deceive us" with a faulty product.[39]

Users of modules in the Gulf of Mexico or anywhere moisture and strong sunshine coexist registered similar complaints. *The Oil and Gas Journal* reported in its March 6, 1978, issue, "Packaging of the cells is a particular concern. Some cells have been packaged in cheap, permeable plastic and have only a 3 to 5 year life expectancy."[40]

Accelerated testing programs conducted by the U.S. Coast Guard and Telecom Australia also exposed flaws inherent in those early panels. But the shortcomings in no way turned people against photovoltaics. Father Verspieren, for example, credited first-generation modules for getting Mali Aqua Viva's solar program started, and he remained confident that photovoltaic systems were the best power source for water pumping in rural Mali, though the technology had a way to go in its development.[41] "Such is the law of life," Father Verspieren stated. "The child takes his first steps holding his father's hand."[42] Others shared Verspieren's tolerance for the problems of the early terrestrial modules. Phillip Wolfe, for example, who began the United Kingdom's first photovoltaics engineering firm by importing Solar Power modules, called them "a fairly crude design." But he added, "Technology has to start somewhere."[43] As Carl Kotiela, a Tideland Signal engineer recalled, when something went wrong, "People didn't get negative toward photovoltaics. They just said we should make a better product."[44] In this spirit, *The Oil and Gas Journal* told its readers, "These low-power applications of solar energy are economical if correctly engineered and debugged."[45] The U.S. Coast Guard, Telecom Australia, Verspieren, and others therefore urged the fledgling photovoltaics industry to better package their cells.

The industry responded quickly with a more durable product. Tideland Signal came up with a more rugged design by introducing a form-fitting molded-glass panel, which completely sealed the cells and their connections from contaminants. But providing protection equivalent to a double glass pie plate had its downside. The module was very bulky and very expensive. And its greater weight and smaller size increased installation costs as well. One veteran called the Tideland strategy, "The total iron-clad, heavy-duty, costs-be-damned approach."[46] Another old-timer described the module as "magnificent, but costly."[47]

Tideland's expensive design was fine for applications such as navigation aids, where solar cells at almost any price would prove economical. But for more garden-variety needs, such as water pumping, a cheaper way to make the panels impermeable had to be found. Photovoltaics pioneer Bill Yerkes came up with the solution. Yerkes had been president of Spectrolab, a leading manufacturer of solar cells for the space industry. When Hughes Aircraft bought the company in 1975, however, he lost his position. Smarting from this abrupt firing, Yerkes vowed, "I'd show them by making a solar company that would be bigger than Spectrolab."[48]

Before opening the doors to his new terrestrial photovoltaics company, Solar Technology International, Yerkes made several important manufacturing decisions. First, he avoided silicone: Its fumes were a problem for workers; it cost a lot; and dirt sticks to it. "People would have to come around every so often and wash the plastic panels with soap and water," Yerkes complained. "On a large scale, this would make the technology unworkable."[49]

He also believed that for photovoltaics to grow into a significant industry, it had to work with low-cost, readily available, long-lived materials that lent themselves to mass production. Based on these prerequisites, tempered glass seemed the best choice for a top cover. It was robust, it was manufactured everywhere, and it would self-clean after a rain.

Finding an effective way to adhere cells to tempered glass did not come easily. In his first attempts to use an acrylic adhesive, Yerkes found that air pockets remained between the cells and the glass. Exposed to the elements, water would condense in these gaps and eventually corrode the metal contacts. A successful product would require a better glue.

While searching for a higher quality adhesive, Yerkes recalled seeing automobiles at wrecking yards: Their bodies were rusted out, but their windshields still looked great. Even after decades in the sun, no air bubbles

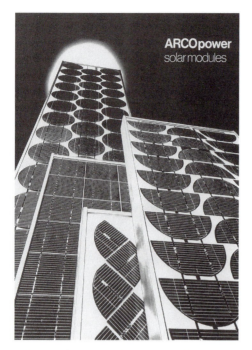

*ARCO Solar utilized techniques developed by Bill Yerkes and his colleagues to introduce relatively low-cost, well-protected solar modules to the industry.*

formed in the composite of two layers of glass bound together by plastic film. Here was a time-tested process that would bond cells to tempered glass—and Yerkes could buy the adhesive by the trainload.

In his quest to streamline production, Yerkes and several colleagues also came up with a more commercial way to apply contacts onto cells. They replaced the labor-, material-, and time-intensive approach of chemical plating with screen printing, the method used to put designs on tee-shirts. Arthur Rudin, who worked with Yerkes, explained the procedure. "You have a screen, you put your shirt under it, the screen has a pattern, and you move a squeegee with paint across the screen. In our case, the paint happens to be a mix of silver paste with glass. The concoction is a good conductor and bonds well with solar cells."[50]

Yerkes took one of his new modules to a meeting at Jet Propulsion Laboratory. To prove its strength, he balanced the panel between two chairs and stepped onto the middle. It did not break. Then he dashed a coffee cup against the module. The cup shattered, but, again, the panel remained intact. Yerkes then told his impressed audience, "These are strong and robust products and they should be since they are going out to work under the harshest conditions."[51]

Soon most photovoltaics manufacturers adopted Yerkes' processes, producing modules of demonstrated durability. A sharp price drop accompanied these improvements, from $11 per watt in 1980 to $7 per watt by 1985, thanks largely to Yerkes' innovations. His company, which was bought by ARCO and is now owned by Siemens, remains one of the largest photovoltaics companies in the world.

The improvements in the manufacture of photovoltaics answered Father Verspieren's prayers. All the modules installed in Mali since 1979, no matter the manufacturer—France Photon, Photowatt, ARCO Solar, or Solarex—have, with few exceptions, performed even better than the extremely reliable immersed motor-pump sets. This caused Verspieren to boast, "By 1981 the solar family at Mali Aqua Viva has become quite numerous and very healthy."[52] Most had functioned without a single problem for more than ten years when Jerome Billerey wrote his 1990 landmark study of photovoltaic water pumping in Mali.[53]

The success of photovoltaic technology, as showcased by the Mali Aqua Viva experience, has created a new problem for solar engineers. The almost flawless performance of the equipment leaves maintenance crews idle much of the time. Mali Aqua Viva's repair crews could service four hundred pumps without being overworked, although presently there are only ninety-six installations under its care.[54] Yet the crews must remain on duty because the sophisticated electronics that have made photovoltaic water pumping so reliable do require the intervention of a specialist should something go wrong.

Mali Aqua Viva has become the leader in photovoltaic water pumping in Africa, if not the world. It has shown that with excellent equipment, good and punctual maintenance, and user participation, photovoltaic-powered water pumping works. An evaluation of Mali's program proves that photovoltaic pumps are more economical than diesel pumps in providing water for human and livestock consumption in small- and medium-sized villages and that they provide more water than manual pumps.[55] From the first few pumps installed by Mali Aqua Viva some twenty years ago, more than one thousand have sprouted in West Africa alone! Over twenty times that number presently serve communities throughout the world. And Father Verspieren's pioneering work deserves much of the credit. "In fact, we can say, 'Thank you, Father Verspieren,' that today we have lots of photovoltaic pumps everywhere," Guy Oliver stated without exaggeration, "and for showing the international community that photovoltaics is an excellent power source for the people of Africa and the rest of the developing world."[56]

The benefits of photovoltaic water pumping carry far more importance than the number of installations. The women of Mali regard solar pumps as one of life's necessities, something they refuse to do without. Before Mali Aqua Viva came along, they had seen their mothers slave, lugging heavy water containers from afar. They will not fall to such drudgery, they say, and therefore swear not to marry a man whose village does not have a photovoltaic-powered water pump.[57] The sight of Malians awash in water has given Verspieren much pleasure, too. "I went as a missionary to convert Africa," he relates, "and one day I met a young woman who gave me water that she had carried for over six kilometers [four miles]. Such generosity of the Africans has converted me and now it is with great joy that I can offer the same woman, who had been resigned to the drought and its terrible consequences, a solar pump and a reservoir full of water."[58]

Verspieren's solar pumping program has had a profound effect on visiting foreign aid workers, too. After viewing one of the priest's first photovoltaic installations, Bernard McNelis, a solar engineer, bolstered his commitment to developing photovoltaic systems throughout the world. The contrast between the healthy, happy Mali Aqua Viva clients who enjoyed a clean and reliable supply of water and the villages that lacked such access, where the inhabitants seemed demoralized and many were debilitated, left a lasting impression on McNelis. "It was amazing to see" how a technology could change the entire tenor of a people, he recalled.[59]

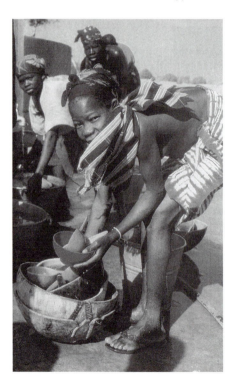

Photovoltaic water pumping, in Verspieren's opinion, is the marriage of the century—uniting water and the sun. There is no chance of divorce, the priest correctly predicted more than sixteen years ago. It is a marriage that becomes more solid over time.

*Washing dishes with solar-pumped water. The sun that brings punishing droughts now also brings life-enhancing water.*

## Notes & Comments

1. M. Chaudey, "Retour au Mali," *La Vie,* 50eme anniversaire (1995).

2. P. Sorelle, "Mali Aqua Viva, A Miracle in the Sahel," *Missionaries of Africa Report,* January/February 1979.

3. B. Verspieren, *Mali Aqua Viva Rapport no. 8—Annees 1982 a 1987,* 48; B. Verspieren, "The Application of Photovoltaics to Water Pumping and Irrigation in Africa," (in French), *Third E.C. Photovoltaic Solar Energy Conference* (Cannes, France) (Dordrecht: Kluwer Academic Publishers, 1981), 442.

4. Sorelle, "Mali Aqua Viva, A Miracle in the Sahel."

5. Verspieren, "The Application of Photovoltaics to Water Pumping," 439; B. Verspieren, *Mali Aqua Viva, Rapport no. 7—Annees a 1979 a 1981*, 25.

6. B. Verspieren, "Mali: un drole de Pere blanc," *Le Point* 775 (27 Juillet 1987): 93.

7. I. Usmani, 1977, "The Challenge of Energy," in N. Veziroglu, ed., *Alternative Sources of Energy: International Compendium, Solar Energy,* vol. 1 (Washington, D.C.: Hemisphere Publishing, 1978), 18.

8. P. Masson, "Report on the Use of Solar Energy for Water Pumping in Arid Areas," 2.

9. H. Tabor, quoted in B. McNelis and A. Derrick, 1990, "Solar Water Pumping—Clean Water for Rural Areas," *2nd ASEAN Science and Technology Week* (30 January–4 February), ASEAN Committee on Science and Technology: Department of Sciences and Technology, Phillipines (1990): reprint.

10. Verspieren, "Mali: un drole de Pere blanc."

11. A. Agarwal et al., *Competition and Collaboration in Renewable Energy* (Washington, D.C.: International Institute for Environment and Development, 1983), 3–6.

12. Interview with Terry Hart.

13. Interview with Dominique Campana.

14. Interview with Wolfgang Palz.

15. D. Campana et al., "Realisation of Testing a Pumping System Powered by Solar Cells," *Proceedings of the UNESCO/WMO Symposium* (Geneva, 30 August–3 September 1976) (Geneva: Secretariat of the World Meteorological Organization, 1977), 612.

16. Ibid., 617.

17. Verspieren, "Mali: un drole de Pere blanc"; Verspieren, *Mali Aqua Viva Rapport no.8*, 17.

18. Campana et al., "Realisation of Testing a Pumping System," 617; Sorelle, "Mali Aqua Viva, A Miracle in the Sahel."

19. Verspieren, 'The Application of Photovoltaics to Water Pumping," 442.

20. Sorelle, "Mali Aqua Viva, A Miracle in the Sahel."

21. Interview with Bob Johnson.

22. Sorelle, "Mali Aqua Viva, A Miracle in the Sahel."

23. Interview with Guy Oliver.

24. "Father Verspieren's Inaugural Speech at Nabasso," *Missionaries of Africa Report* (January/February 1979).

25. Ibid.; Verspieren, "Mali: un drole de Pere blanc," 93; Sorelle, "Mali Aqua Viva, A Miracle in the Sahel."

26. Verspieren, "Mali: un drole de Pere blanc," 31.

27. *Mali: Solar Energy Country Report* (McClean, VA: PRC Energy Analyses Company, 1979), 4.

28. Verspieren, "The Application of Photovoltaics to Water Pumping," 441.

29. Agence Française pour la Maîtrise de l'Energie et al., *Le Pompage Solaire Photovoltaique: 13 annees d'experiences et de savoire faire au Mali*, Annexe (Paris: Agence Française pour la Maîtrise de l'Energie, 1991), C1, C2; A. Haentijens, *Presentation et evaluation du programme "energies renouvelables" mise en ouevre par Mali Aqua Viva* (Paris: Agence Française pour la Maîtrise de l'Energie, 1984), 16.

30. A. Haentijens, *Presentation et evaluation du programme*, 17.

31. Verspieren, *Mali Aqua Viva Rapport no. 7*, 23.

32. Ibid.

33. Interview with Momadov Diarra.

34. Haentijens, *Presentation et evaluation*, 27–28.

35. Interview with Terry Hart.

36. Verspieren, "The Application of Photovoltaics to Water Pumping," 439.

37. Direction Nationale de l'Hydraulique et de l'Energies, Republique du Mali, Cellule d'Entretien des Equipements Solaires, *Le Pompage Solaire Photovoltaique au Mali 1977/1990* (Bamako, Mali) (1990): 15; Haentijens, *Presentation et Evaluation*, 28.

38. Agence Française pour la Maîtrise de l'Energie, *Le Pompage Solaire Photovoltaique*, 10.

39. Verspieren, "The Application of Photovoltaics," 439.

40. D. Nolan, "Solar Energy Used for Production Applications," *The Oil and Gas Journal* 76 (6 March 1978): 88.

41. Verspieren, "The Application of Photovoltaics," 439.

42. Verspieren, *Mali Aqua Viva Rapport no. 7*, 24.

43. Interview with Phillip Wolff.

44. Interview with Carl Kotiela.

45. Nolan, "Solar Energy Used for Production Applications."

46. Interview with a Photocomm executive.

47. Interview with Michael Mack.

48. Interview with Bill Yerkes.

49. Ibid.

50. Interview with Arthur Rudin. Bill Yerkes began shipping solar cells with screened interconnections as early as January 1976, long before any other photovoltaics manufacturer put the process to use. L. Curran, "Companies Look for Ways to Raise Solar-Cell Output," *Electronics* (11 November 1976): 94.

51. Interview with Bill Yerkes.

52. Verspieren, *Mali Aqua Viva Rapport no. 7*, 23.

53. Agence Française pour la Maîtrise de l'Energie, *Le Pompage Solaire Photovoltaique*, 17.

54. Direction Nationale de l'Hydraulique et de l'Energies, *Le Pompage Solaire.*

55. J. Billerey, private communication.

56. Interview with Guy Oliver.

57. Verspieren, "Mali: un drole de Pere blanc." The Malian women do not exaggerate. Nyamutondo Gaspari of Mwanza, Tanzania, attributes her chronic leg and lower back pain to carrying a four-gallon bucketful of water a half mile from Lake Victoria to her home day after day. M. Plummer, "To Fetch a Pail of Water," *Natural History* (February 1999): 56.

58. Verspieren, *Mali Aqua Viva Rapport no. 8*, 442. For the most thorough study in English of Father's Verspieren's work in Mali, consult Meridian Corporation and IT Power Ltd., *Learning from Success: Photovoltaic-Powered Water Pumping in Mali* (20 February 1990).

59. Interview with Bernard McNelis. Father Verspieren's early accomplishments had a significant effect on Bernard McNelis' career in renewable energies. It stimulated his interest in the potential of photovoltaics for water pumping and led him and his firm, the Intermediate Technology Development Group, which later became IT Power, to participate in the first comprehensive studies on photovoltaics for water pumping. These studies facilitated the development of "small photovoltaic pumps . . . to the stage where the best can meet all the technical and user prerequisites for wide-scale introduction." B. McNelis, "Photovoltaic Water Pumping—A 1986 Update," *Seventh E.C. Photovoltaic Solar Energy Conference* (Seville, Spain, 27–31 October 1986) (Dordrecht: Kluwer Academic Publishers, 1987), 27.

Conversely, McNelis' early recognition of Verspieren's work in international publications brought greater notice to Mali Aqua Viva's pioneering work, thus helping the organization realize its ambitious goals.

## Chapter Twelve
# Electrifying the Unelectrified

Cold war warriors in both the Soviet and American camps saw nuclear power as the key to winning the hearts and minds of the developing world, believing that the electricity it generated would "enable [such] countries to reach the standards of living of the industrialized countries."[1] Even those suggesting a more diverse power mix to electrify villages, where most of the developing world's population resides, still assumed wiring them to some sort of central power station.[2] While planners dickered over the means of generating electricity, the cost of stringing the wires to deliver the energy did not figure into their discussions. As it has turned out, even if energy from a power station were "too cheap to meter," no developing nation could afford to deliver it to outlying communities. That is because, as one knowledgeable writer explained, "It's one thing to build a 180 [megawatt] dam and march the power into the city with high-tension cables, it's another to distribute power to the [majority] of the population."[3] The cost of hardwiring the countryside is just too great. Power line extensions can cost tens of thousands of dollars per mile. In the developing world, it is unlikely that a utility would ever be reimbursed for such an outlay; customers could never pay back the investment. Rural consumers in poorer countries do not use that much electricity because they can only afford to buy a light bulb or two and perhaps a radio or tape cassette or television.

Either they would have to pay way beyond their means for each kilowatt hour they use or the line extensions would have to be subsidized by governments already mired in debt. For these reasons, one-third of humanity, over two billion people, live without electricity. It has now become apparent to many energy planners that the "obvious classic approaches will not bring more and better energy" to these people. That is why "we are looking for alternatives," Rob Van der Plas of the World Bank recently stated.[4]

Stand-alone power systems have become the poorer nations' only hope. Diesel generators were once considered the answer, but their continued functioning relies largely on fuel trucks arriving on schedule. During the rainy season, the trucks often show up late, if at all. Even in the best weather, roads pitted with potholes, if paved, make broken axles common, and, again, the fuel doesn't get delivered.

The logistics of transporting fuel to generators located in remote sites can border on the absurd. Trucking hydrocarbons to small outposts in the Amazon, for example, consumes two to three gallons of diesel fuel for every gallon a generator uses! The high price of diesel fuel, due to the expense of transportation and sharp rises in petroleum costs since the early 1970s, has forced most owners to drastically curtail the number of hours they run their generators. And should breakdowns occur, as they always do, it may take weeks or months before repairs can be made because parts and skilled mechanics are few and far between. Such problems have led many energy experts to conclude that the diesel engine is "a less elegant solution than one would think" for rural electrification.[5]

The hundreds of millions of people in the developing world who do not own diesel generators spend around $20 billion a year on ad hoc solutions such as kerosene lamps, candles, or even open fires for lighting, and batteries for radios, TVs, and cassette players. But none of these options can compare to uninterrupted electricity. A candle, for example, gives off one lumen of light and a simple oil lamp provides ten lumens, while a ten-watt fluorescent tube provides five hundred lumens.[6] For a store owner, the difference in illumination creates a complete change in business. With kerosene lights, "The inside of the store is dark and clients hesitate entering," a shopkeeper in the Dominican Republic states. "But with good lighting, everyone comes in and spends their money."[7] Like the fuel for generators, kerosene and batteries come from outside suppliers. When the local store owners run out, everything goes dark.

For reliable power, unelectrified rural people must find an indigenous energy source. For the few who live where a river flows year round or the wind blows steadily through every season, water or wind power might serve them best. The sun, however, shines everywhere. A residential photovoltaic system takes less than one day to get up and running, costs a few hundred dollars, and, when connected to a battery to store electricity for sunless periods, operates as a self-sufficient unit day and night on a readily available fuel. In contrast, nuclear power stations and coal- or oil-driven electric plants require huge capital expenditures, years of construction, imported fuel, and transmission lines. Hence, over time strong support has developed for photovoltaics as the best choice for most rural electrification programs.

Many who advocated photovoltaics as a viable energy source for the developing world first considered outfitting villages with miniaturized centralized power stations. For example, in 1977, one United Nations energy expert proposed the installation of ten thousand fifty-kilowatt photovoltaic plants to supply all the electrical needs of ten thousand villages housing one hundred families each.[8] A bank of panels would be set up close to the village, along with auxiliary equipment, such as batteries. Wires would run from the solar-generating plant to the nearby homes.

Multi-kilowatt photovoltaic installations did not work as their advocates expected. A case in point was the photovoltaic mini-plant at a village on the South Pacific island of Utirik. The problem was that each household tried to get as much power from the system as it could, burning lights all night and running oversized appliances. Homeowners not officially connected to the system would surreptitiously tap the power lines. The excessive—and unanticipated—demand caused the power plant to constantly crash, leaving no one with electricity. Jim Martz, the NASA engineer who troubleshot the project, soberly concluded, "There would have to be someone there all the time who could watch over it and make sure no one cheated. In other words, it had to be policed to operate."[9]

Imposing such measures seemed too draconian. Therefore officials on the island decided to divide up the panels and batteries and hand them out to each villager. Experience at Utirik and other locales has proven that photovoltaic panels placed on individual homes have a greater chance for success than photovoltaic mini-power plants "because families invariably consume more when they are not personally responsible for their [photovoltaic] system and [their] consumption" of electricity.[10] Furthermore, if an individual overloads a system and wrecks it, it is that person's problem, not the village's.

The use of photovoltaics for individual remote homes in the developing world was pioneered by the French. Ironically, it was the French Atomic Energy Commission that initiated the program in 1978. The agency's nuclear testing in Polynesia had not endeared it or the French government to the Polynesian people. Public opinion had to be shored up. "We wanted to be popular," Patrick Jourde, who worked for the commission in Polynesia at the time, admitted. "We wanted to be known not only for nuclear tests but also for helping out the people."[11] Bernard McNelis, a British colleague of Jourde's, described the agency's intentions more bluntly: "The original motivation for the program was a bit doubtful because it had to do with justifying the French nuclear presence. The attitude was, give the people electricity to keep them happy."[12]

Jourde was in charge of finding a way to bring electricity to those who lived on the isolated atolls and islands scattered over the South Pacific Ocean. The distances between each atoll and island, and the small populations that lived on them, forced Jourde to look for new ways to provide electricity to them. He experimented with various renewable technologies that could use locally available fuel. Jourde first considered gasification, burning coconut shells and the like, but the difficulty of collecting sufficient quantities ruled that out. Wind was also considered, but high maintenance and the considerable variability in supply doomed this choice. Photovoltaics alone seemed to fit the bill. "With one week of storage [capacity]," Jourde explained, "you have complete autonomy."[13]

Having made the decision to electrify the outer islands with photovoltaics, Jourde first helped establish a for-profit business, G.I.E. Soler, to assemble the components necessary for a solar electric house to function smoothly, including the design and manufacture of energy-efficient appliances specifically for photovoltaic-powered homes. Because they consumed considerably less energy, these appliances minimized the number of panels—the most expensive item in the system—that each house needed, without compromising comfort. Jourde also organized a photovoltaics information, research, and training center to give customers excellent service.

When the infrastructure was in place, he targeted sales to the most respected people. One of his first customers was Marlon Brando. "Everyone else followed their example," said Jourde. The affluent purchased four hundred-watt panels that would run lights, a TV, hi-fi, refrigerator, washing

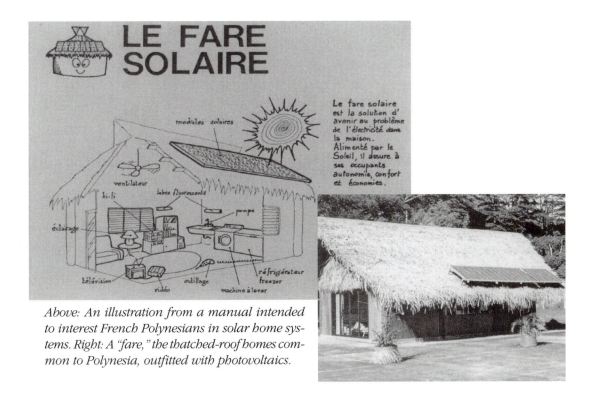

*Above: An illustration from a manual intended to interest French Polynesians in solar home systems. Right: A "fare," the thatched-roof homes common to Polynesia, outfitted with photovoltaics.*

machine, fans, and power tools.[14] Together the appliances and photovoltaic system cost around $10,000. Those with less money could buy a smaller package for approximately $2,000. A 20 percent subsidy and a low-interest loan payable over five years made such purchases more appealing. Still, considering "the other choices—for example, a diesel generator—the photovoltaics option was cheaper than any other form of energy," Jourde attested.[15]

Between 1980 and 1990, G.I.E. Soler installed photovoltaic systems on thirty-three hundred houses. By 1987, one-half of all electrified homes in French Polynesia ran on power from the sun. Of these installations, 50 percent were purchased by homeowners who continue to add panels when money permits. For the other 50 percent, the poorer islanders, photovoltaic systems were installed for free with the understanding that they would pay a monthly fee to the mayor of their village. This would reimburse the municipality, which had purchased the modules from G.I.E. Soler and had paid for their installation. But once the solar was in, many refused to hand over a centime. "Look," they told their mayors, "we do not pay for water, do we? That's because it comes in raindrops sent by God. Doesn't God also send down the sun's rays which energize our solar panels? Then why should we pay for electricity from the sun?" The mayors, who relied on their constituency for reelection, did not refute such logic. And G.I.E. Soler had no other way to collect.

The people who had not paid for their systems tried to get as much energy as they could from them, caring little if the batteries were totally discharged and ruined. They knew that G.I.E. Soler had to fix whatever went wrong. It ended "up like the public health system in France: the users [ran] riot" with the solar equipment, Jourde stated.[16] From this experience, Jourde learned two valuable lessons: giveaway programs invariably fail, and only when the battery is working well does a photovoltaic system function properly. Or in the words of a battery specialist with considerable experience in photovoltaics, "The battery is the heart of any stand-alone installation. If you don't have a reliable battery, you don't have a working system as you are essentially using the solar cells for a battery charger."[17]

After the batteries had been ruined, people hooked appliances directly to the modules. On Makatae Island, far from Tahiti, the islanders had done just this. Jourde arrived there very early one morning. As the sun rose and its rays bore down on the cells, music from forty huge stereos became audible. At midday, the music boomed all over the island, dying down only as the sun set. "It was quite funny," Jourde recalled. "Everyone was very happy" with the way the stereos worked. "It seems that lighting was not their first priority!"[18]

In some instances, when the local utility wired a village formerly powered solely by the sun to a diesel-run power station, all trouble ended. But no one tore down their photovoltaic modules  because the two means of power complement each other. As in most of the developing world, electricity generated by diesel comes only in the morning and evening. Generators are run intermittently to save fuel and minimize wear and tear. The solar cells then provide enough charge to carry through when diesel-generated electricity is not available. As Jourde observed, "People are happy to have a little diesel and a little sun power."[19] With this combination, they have electricity pretty much all the time.

Nearly 20 percent of the world's production of solar cells in 1983 found its way to the thatched roofs of French Polynesia.[20] Elsewhere, neither government nor private industry seemed to have a clue about how to successfully electrify rural homes in the developing world using photovoltaics.[21] No one matched the work done by Jourde and his staff. Dealing with a dispersed market that needed very small increments of power—the situation where photovoltaics will always outperform its rivals—proved beyond their ken. For example, in 1984, the Ceylon Electrical Board, Sri Lanka's national utility, offered solar cell home power kits to its unelectrified rural constituency at cost. However, no bureaucrat had considered traveling from village to village to demonstrate their use. The utility placed advertisements in major newspapers informing villagers that they had to come to its head office in Colombo, the nation's capital, to pick up the kits. After purchasing them, the buyers were on their own. The utility did not offer any help in installation or upkeep. Not surprisingly, only six hundred kits have been sold over the years.[22]

Nor did the ARCO Solar dealer, the largest distributor of photovoltaics in Sri Lanka in the mid-1980s, display any interest in selling solar door-to-door. He felt more comfortable remaining in Colombo and arranging large contracts with the government, selling modules for telecommunications and navigation aids. Despite the fact that the vast majority of Sri Lankans had no electricity, he dismissed the possibility of doing business with villagers; he believed that those living in "the jungle" had no cash.[23]

A similar attitude prevailed in Nairobi, Kenya. Firms representing the major manufacturers had set up shop in the nation's capital in the late 1970s and early 1980s to take advantage of the growing demand for photovoltaics generated by Nairobi-based international aid organizations serv-

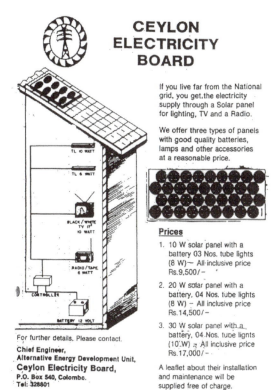

*In 1984, advertisements like this one appeared in the major Sri Lankan newspapers to inform villages that the Ceylon Electricity Board, the national utility, was offering low-cost solar electric kits for sale. The Board, however, did little beyond placing the ads and the program, therefore, went nowhere.*

ing eastern Africa. These groups had found that solar cells were the only power source they could depend on to run vaccine refrigerators, water pumps, and other electrical equipment outside the towns and cities, and to electrify fences in wildlife reserves to keep game animals from wandering into farms. With a brisk business in the capital, no one saw any reason to leave the comforts of the city to drum up trade in the countryside. Yet there were millions of middle-class rural Kenyans, like the Mugambis, who taught school in the countryside. They had wanted electricity for the longest time, but had given up hope of ever connecting to the power lines. "If we have to wait for the government [the sole supplier of utility-produced electricity in Kenya]," Mrs. Mugambi complained, "we will be old and grey."[24] Indeed, Kenya Power and Lighting's service ends more or less at the city limits, leaving almost all rural Kenyans without electrical hook-ups. By 1996, the utility's rural electrification program, started in 1976, had reached only about thirty thousand of the millions living in the countryside. "At this pace," one commentator observed, "it will take more than eight hundred years before every rural household in Kenya has been connected."[25]

People like the Mugambis want the amenities that run on electricity, such as TV and good lighting—and they have the means to purchase them. Some move to the cities in order to have these conveniences. Others, however, such as Joseph Omokambo, a Kenyan civil servant, have stayed put but still manage to enjoy TV. Though not connected to a power line, Omokambo found that he could get the electricity he needed by hooking a car battery to his television set. Once or more a week, depending on how much TV he watched, the battery would run out. Recharging meant lugging the battery three or so miles (five kilometers) to town and leaving it overnight with an enterprising townsman who had electricity and who had set up a makeshift charging station in his garage. Omokambo would return the next morning, pay the man about $2, and carry the battery home. As time went on and the battery wore down, the trips became more frequent. Although he considered it burdensome, Omokambo continued to cart the battery back and forth. Otherwise, he believed, he would have to give up television. Then he learned from friends that there was a way he could keep his battery charged at home. Solar Shamba, they told him, sold a gadget that when placed on the roof produced electricity.

Solar Shamba's founder, Harold Burris, was an engineer by trade who had come to Kenya with the Peace Corps in 1977. He later married a

Kenyan seamstress. When his wife complained that her foot-powered sewing machine stitched too slowly, he connected it to a DC motor powered by a photovoltaic panel. This, he surmised, would not only solve his wife's problem, but its mass use would better the lives of many Kenyan women. Despite zealous attempts to get The Singer Company involved and to sell the idea to treadle machine owners, neither shared Burris' enthusiasm.

In 1984, at about the same time that Burris established his small solar enterprise in the prosperous coffee-growing region near the slopes of Mount Kenya, a nearby rural boarding school decided to replace its kerosene lanterns with electric lights. The school's board of governors considered two options: It could hook up the school to Kenya Power and Lighting's lines or it could buy a generator. Since the school was over four miles (6.5 kilometers) from the nearest utility pole, a connection would cost about $21,000—far too high a price for the tiny school's limited budget.[26] Therefore, the board decided on the diesel generator.

A few days before the purchase order was placed, the school's Peace Corps science teacher, Mark Hankins, suggested that the administrators consider photovoltaics instead. When the board countered with, "It's so expensive," Hankins presented a comparative economic analysis that showed that a photovoltaic system was cheaper than diesel.[27] The board remained skeptical, so Hankins suggested that the members visit Burris' photovoltaic-powered office/home and see for themselves. Except when a generator made its presence known by its noise and odor, these rural educators had never seen such brightness after dark.

The demonstration at the Burris home led the school to postpone its purchase of the generator and to try solar lighting in four classrooms,

*Harold Burris (with beard), founder of Solar Shamba, and locally trained solar installers prepare to put up a photovoltaic module.*

*Solar electricity allows this Kenyan family to read without the fumes, fuss, and eyestrain that come with kerosene lighting.*

using locally built batteries and lights. Only the module was imported. Three months later, the system was up and running. The students' eyes no longer teared from the smoke of the old kerosene lanterns nor strained under poor illumination. They no longer had to crowd around the lamp like so many moths for enough light to read.

The trial run went so well that the school administrators scrapped the generator idea and lit the laboratory, administration office, the remaining classrooms, and four teachers' homes with solar modules.[28] Visiting educators, once they saw how well illuminated the school was, requested solar electricity for their institutions. Nighttime passersby could not help but notice, and the more curious stopped to inquire about the new technology.

Word spread about Burris' solar electric generator. Rural people with disposable income—successful farmers, civil servants, doctors, teachers, and other professionals—who wanted a convenient way to run their TVs, radios, cassette players, and lights made their way to his shop. Burris sensed where the market was heading and altered his course to satisfy the demand. He designed a modular thirty-five-watt residential solar system that would provide enough energy to satisfy the people's electrical needs. Assisted by the U.S. Agency for International Development (U.S. AID) small project grants, Burris and

*Mark Hankins (left), who helped to popularize solar home systems in Kenya, measures the voltage of a module with a workshop participant.*

Hankins held a photovoltaics installation training course for a dozen local electricians, who then became his sales- and workforce. In fewer than five years, Solar Shamba had electrified more than five hundred homes, mostly in the Meru district where the American engineer lived.[29]

Burris sent his technicians to Nairobi to buy the panels. Once they knew where to get them and had installed several systems under his direction, many felt confident enough to go into business for themselves. As sales increased, the Nairobi-based photovoltaics suppliers began to enter the residential market and sell aggressively to households, too. Currently there are scores of companies in Kenya selling solar home systems in almost every town. Photovoltaics has grown into a multimillion-dollar business, with sales increasing 25 to 30 percent per annum over the last seven years.

A very small module has come to dominate the market. Rated at only twelve watts, and priced at about $80 U.S., it is affordable for many Kenyans. Over ten thousand twelve-watt units have been purchased each year since 1994. The mini-module kit is sold alongside televisions and radios, because "shopkeepers have realized that if people buy a TV, they need something to power it," observed Mark Hankins, still at the forefront of the Kenyan photovoltaics scene. He has discovered that the majority of small system

*Daniel Kithokoi began his solar career working for Harold Burris. He later started his own successful photovoltaics business at the foot of Mt. Kenya.*

*A typical rural photovoltaic installation in Kenya. The module is extremely small compared to those typically used in the developed world.*

owners "learn through experience that with such a tiny power source they can't run their lights and TV all the time. So, they learn to ration their power needs."[30]

Two-thirds of all home photovoltaic systems in Kenya work as they should. Another 13 percent are partly functional—the TV still goes on, but some lights don't. Placed in the context of a country where the traffic lights in the capital go out and stay out for months at a time and where central library researchers recommend packing a flashlight because the lights don't always work, photovoltaics' track record appears exemplary. This is especially impressive considering that many installations are done ad hoc and in very remote areas.

After interviewing a number of householders who successfully power their homes with solar electricity, Richard Acker, who conducted a solar survey for Princeton University, found, "These people, who are the majority of photovoltaics users, are very enthused because it has taken them a big step closer to enjoying the comforts promised by modern technology. They now have a reliable source of electricity."[31] They do share one complaint, however: They all would like to squeeze more power out of their modules, especially during the rainy season.

Neighbors of photovoltaics owners almost unanimously would buy solar modules if they could, though they may share Mrs. Mugambi's initial reservations: "At first, when my husband launched the idea of buying a [photovoltaic] system, I thought it was too expensive for us [as it cost] about three months of both of our salaries." But after calculating the amount spent on kerosene and rechargeable and throwaway batteries, many Kenyans have come to the same conclusion as the Mugambis—that they "already spent quite a lot of money on energy" and that the purchase of a photovoltaic system is a good investment.[32]

Of the estimated four hundred thousand home solar systems in the

*An advertisement for solar electricity in a Kenyan farm town. Note the ad for kerosene lamps, the technology solar replaces, on the right.*

world, sixty thousand are in Kenya.[33] With 2 percent of the rural populace relying on solar power for their electricity, Kenya has become the first country where more people plug into the sun than into the national rural electrification program. What is more amazing is that photovoltaics' ascendancy has occurred without government help. The Kenyan experience suggests that the principal power source in rural Africa is going to be photovoltaics.

While the Kenyan market for solar electricity continues to expand, having to pay for a system in one lump sum puts a damper on many consumers' enthusiasm, and it limits the size of a system they can afford. Paying up front essentially forces people to buy thirty years of electricity all at once. Richard Hansen, an American engineer and MBA who promotes photovoltaics in the Dominican Republic, is not alone in viewing financing as the primary obstacle to the technology's widespread acceptance.

Though he worked for Westinghouse, Hansen's true interest was in renewable energies, having studied wind power at Worcester Polytechnic Institute. While traveling around the Dominican countryside on vacation, he began to consider the possibility of making a living selling small wind machines on the island, since no "investments in the infrastructure of conventional power systems in the rural areas had been made, and people were attempting to get increasing amounts of energy."[34]

When his desire to strike out on his own finally overwhelmed his need for security, Hansen quit his job. Being a mechanical engineer, he preferred wind machines to photovoltaics, admitting a certain repugnance to a device "that just sits there and converts sunlight to electricity." But Hansen soon learned that the wind blows steadily in only limited areas of the Dominican Republic, while the sun shines with the same intensity almost everywhere on the island. He also discovered that the mechanics of windmills, which he loved, proved to be a liability since moving parts tend to break down. Photovoltaics, Hansen observed, "had nothing that spun or moved and therefore would never wear out." Though it pained him to put wind machines aside, his studies convinced him that for Dominican households solar cells "offered greater potential."[35]

He then studied the energy habits of those living in the countryside. He found that most Dominicans follow the news, so they have radios; many love music, so they own cassette players. A simple photovoltaic system could easily power these devices, Hansen surmised.[36] So he set up a demonstration system at the house he had rented. Guests invited to visit on Sunday, the day Dominicans socialize, marveled at the TV, radio, and lights powered by the sun. Many were interested in buying a system, but no one had enough cash to purchase it outright. Because his own funds were limited, Hansen could only sell one system on credit at a time. He chose as his first customer the Martinez family, who owned a market in a heavily frequented area. "People would come by and see the system at work," Hansen recalled. "The family was very friendly and turned out to be very good promoters because of their outgoing manner. As a result, they stirred up a lot of future business."[37]

Back in the States for a visit, Hansen received an urgent letter from the Dominican Republic. Nothing was wrong with the system, the Martinezes wrote, it was performing "exactly as you told us it would, Richard." The problem was that "many are waiting for you to return to negotiate the purchase of these solar systems."[38]

The Martinez family's ability to meet payments convinced Hansen that rural Dominicans could buy solar home systems—if the cost could be spread over a number of years. With financing, "the technology would become affordable to many."[39]

Returning to the Dominican Republic, Hansen met with his prospective customers. Together they formed the Dominican Families for Solar Energy, which came up with the idea of a revolving fund. Seed money would provide loans for an initial set of installations; the money collected from those carrying the loans would pay for additional solar systems. A $2,000 grant from U.S. AID allowed five members of the Dominican Families for Solar Energy to solar electrify their homes in March 1985. The $50 down payment and $8 monthly installment that each family made provided the funds for a sixth family to get a photovoltaic system that June.[40]

While the cash provided by the revolving fund was better than nothing, Hansen and his Dominican clients could see its limi-

Local Dominican solar technicians installing a module. The Spanish reads: "We bring light to the countryside. Rural areas ought to be made more habitable for the good of all the country."

tations. The U.S. AID seed capital would electrify only twenty homes in five years.[41] Even if that amount were raised to $100,000, merely eight hundred solar-powered homes would be up and running in that same time period.[42] As one of Hansen's co-workers lamented, "The demand for loans for solar electricity far exceeds the capacity of the fund." At one point it was "not possible to install anything for lack of money," the president of the rural solar association complained.[43]

So Hansen devised a new approach. He realized that if the burden of the down payment could be eliminated, the customer base would rise significantly. Through a market survey, he learned that perhaps as many as 50 percent of the unelectrified would choose photovoltaics if they only had to pay a monthly fee equal to the amount they currently paid for energy—and if maintenance were guaranteed.[44]

One way to lower the entrance cost, Hansen figured, would be to sell a photovoltaic service rather than a system. Instead of expecting the customer to purchase and maintain the capital equipment, an investor-owned solar utility could collect a monthly fee for providing reliable solar electricity.

To better understand how investors might effect the successful transition from one energy source to another, Hansen returned home to Massachusetts to study the strategy electric companies had taken to win gas users over to electric lighting at the turn of the last century. By 1993, Hansen and his colleagues felt ready to implement a similar strategy in the Dominican Republic. Hansen's company, Soluz, "puts up the modules and wires the house . . . with the lights included. Customers pay a monthly charge and for that they get a complete service commitment. We replace burnt out bulbs or spent batteries. This is exactly what the electric companies did in the early days," said Hansen. "They gave the lamps away, wired the houses for free. That's how people went from gas to electric."[45] And this is how people in the Dominican Republic and Honduras are making the change from batteries and kerosene to electricity generated by sunlight.

The solar utility concept works for both consumer and investor. "[The consumers] keep their capital," so there's no risk for them, Hansen explained. "Since they don't own the equipment, they'll simply stop paying the fee if the service isn't good. They therefore feel pretty secure that we are going to go out there and make sure that everything is operating." Conversely, those leasing a system tend to take good care of the equipment since the company can easily remove the power unit. No one "wants to lose their service once they have electric lights and have reduced their dependency on dry cells," Hansen observed. "We either get the money or repossess the unit."[46] In addition, investors can look forward to an excellent return on their capital and that prospect attracts more expansion capital.

Soluz Dominicana, Soluz's Dominican subsidiary, has been in business since 1994. Since that time, it has outfitted more than two thousand homes with photovoltaic power, and the company reaches between 50 percent to 90 percent of each community it enters. The heavy density of photovoltaics users eases the cost of fee collection and upkeep. Trained technicians do not have to travel far-flung routes to service widely scattered units, as they did prior to the utility's existence. "Before we were just skimming the cream—the cash and credit business," Hansen recalls. "We had to drive all over the place to find customers who could afford it. Now we electrify most everyone. This helps to economize the operation."[47]

Soluz plans to have twenty thousand customers in its network by the year 2000, with the potential to reach fifty thousand to one hundred thousand over the next five to ten years. The solar utility approach "has more impact on the affordability and market penetration of [photovoltaic] technology than would a major newsworthy 50 percent reduction in the cost of [photovoltaic] cells."[48]

Providing credit at reasonable rates to rural people in developing countries could also enable the unelectrified masses to power their homes with photovoltaics. One company tackling the financing problem in this way is the Solar Electric Light Company (SELCO).

SELCO is the brainchild of Neville Williams, who first learned about photovoltaics while working at the Department of Energy during the Carter years. The drop in both oil prices and interest in solar energy during the Reagan administration cost Williams his job. After years of working in unrelated fields, he found himself wondering whatever happened to solar energy. To satisfy his curiosity, he hired on as a consultant to Solarex in the late 1980s. "I was kind of amazed at what I found out," Williams recalled. Not only had photovoltaics not died, "it was actually growing around the world, though nobody knew this. The World Bank was completely oblivious. United Nations' agencies hadn't a clue."[49] Williams pondered the role he could play in this silent revolution.

One night in 1990, while traveling in Africa, he awoke with the answer. "I thought that I would form an organization that would promote the use of photovoltaics worldwide."[50] His new calling led him to visit Richard Hansen (this was in the days before Soluz). What Hansen had accom-

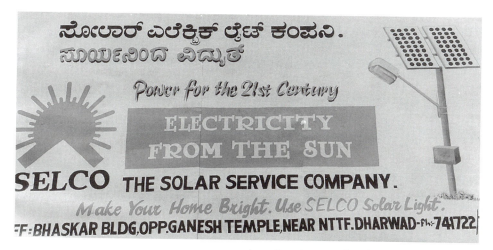

ನೋಲಾರ್ ಎಲೆಕ್ಟ್ರಿಕ್ ಲೈಟ್ ಕಂಪನಿ.
ಸೂರ್ಯನಿಂದ ವಿದ್ಯುತ್

*Power for the 21st Century*

**ELECTRICITY FROM THE SUN**

**SELCO** THE SOLAR SERVICE COMPANY.
*Make Your Home Bright. Use SELCO Solar Light.*
FF: BHASKAR BLDG, OPP. GANESH TEMPLE, NEAR NTTF. DHARWAD-Ph: 741722

*An advertisement for the Solar Electric Light Company (SELCO) in southern India.*

plished with a pittance demonstrated to Williams the power of a well-planned dream. So Williams returned to the United States, formed a non-profit organization, which he named the Solar Electric Light Fund (SELF), and raised a considerable amount of money through fund-raisers and from foundations. SELF closely followed Hansen's blueprint: supply only good equipment; provide maintenance; and develop a repayment scheme and a means of collection. In the process, SELF again proved that people in the developing world do have money and they are willing to spend it on a better energy source—if financing is available. After launching solar rural electrification projects in eleven countries, Williams learned that the demand for photovoltaics far outstrips the credit that foundation grants can provide. He learned the constraints of a nonprofit, just as Hansen had.

Despite the limitations Williams faced, SELF had established enough credibility to attract such people as Harish Hande from India. Hande, who was working on his doctorate in solar engineering at the University of Massachusetts when he walked into SELF's Washington office. The two men agreed that although SELF had demonstrated that there was great interest in solar home systems, only private capital could finance installations on a significant scale. They therefore decided to form a private company, the Solar Electric Light Company, better known as SELCO.

Hande and Williams modeled their start-up photovoltaics company on the U.S. auto industry. The automobile's great success depended on the availability of service and third-party financing. It is unlikely that the masses would have ventured out onto the roadways without access to mechanics and spare parts for their new vehicles. Hande and Williams believed the same applied to photovoltaics. So they set up several solar service centers in South India, which had been chosen for its relative prosperity and need for electricity, to handle sales, coordinate installations, and provide panels, lights, needed accessories, and repairs,

Hande and Williams also realized that auto dealers would have sold few cars had the customer been expected to pay the full sticker price in one payment. How then could anyone expect large numbers of farmers in the developing world to buy a photovoltaic system without financing? To set up credit arrangements in India, they knew they must first impress bankers with their product. SELCO therefore electrified a number of branch offices with photovoltaics. Whereas erratic power supplies had made running a rural bank difficult, the steady power the sun supplies keeps the tellers' computers running throughout the day and gives them reliable lighting at night. "They see that photovoltaics works, that it's loanable, and it then joins a list of products that they will finance," according to Williams.[51]

The World Bank's recent readiness to make significant funds available for financing solar home systems has boosted Williams' hopes of helping to solarize other parts of the world as well. In his estimation, "There's no end to customers if you can make it affordable, with low enough monthly payments."[52] As proof, SELCO has an agreement in Vietnam to electrify one million households within the next decade. It hopes to extend its services to almost three million households worldwide in that time.

*This poster illustrates how photovoltaics will bring electricity to unelectrified Vietnamese homes.*

Perhaps nothing better demonstrates the high value those living in the developing world place on solar cells than the drastic increase in solar module thefts over the last few years. "People steal modules because everyone knows their worth and that they are easy to use," commented Guy Oliver. "Either the modules are resold or they are wired to the thief's battery or directly to his television, radio, or cassette player and it works!" the French engineer added.[53]

In one solar crime, the perpetrator knocked on the door of a British expatriate who had a fairly extensive installation of four forty-watt panels. With a gun to his head, the Englishman watched helplessly as several confederates armed with bolt cutters and crowbars took down the panels. In another incident, a telecommunications contractor wondered why his recently installed system had gone down. An inspection revealed that one hundred modules had been stolen![54] Before the solar home market exploded, a fence would be placed around photovoltaic installations to keep livestock from damaging the panels. Now razor-wire curls around the top of such enclosures to discourage solar thieves.

## Notes & Comments

1. L. Strauss, quoted in *Proceedings of the International Conference on the Peaceful Uses of Atomic Energy*, vol. 16 (Geneva, Switzerland) (New York: United Nations, 1955), 33.
2. D. Smith, *Photovoltaic Power in Less Developed Countries* (Lexington, MA: MIT, Lincoln Laboratory, 1977), 13–14. Elliot Berman was one of the first to recognize the paradox that even if a large central power plant produced electricity "too cheap to meter," rural customers in the developing world still could not afford it because of the expense of delivery. He therefore was also one of the first to see that "because [photovoltaics] can avoid the major problems inherent in other power systems [such as] . . . fuel transportation and power transmission . . . the sun is an attractive energy source" for more than two hundred million households distant from electric lines. E. Berman, "Solar Power" (1968) 1.
3. "Photovoltaic Technology is Taking Root in East Africa," "Solar in Africa," *Ecotec Resources bv* (1994): 5.
4. Interview with Rob van der Plas.
5. "Energy Problems of Developing Countries—West Africa—The Principal Energy Problems of West Africa," *Energy Needs/Expectations* (13th World Energy Conference, Cannes, France) (London: Organizing Committee, 13th World Energy Conference, 1986).
6. M. R. Starr et al., "Design of PV Lighting Systems for Developing Countries," in *Right Light 3, 3rd European Conference on Energy Efficiency Lighting* (Newcastle-upon-Tyne, U.K., 18–21 June 1995), reprint.
7. P. Covell, "Posibilidades Nuevas para la Electricidad Rural," *Desarrollo Suplemento Listin Diario* (27 March 1987): 9.
8. I. Usmani, "The Challenge of Energy," in T. Nejat Veziroglu, ed. *Alternative Sources of Energy: Solar*, vol. 1 (Washington, D.C.: Hemisphere Publishing, 1977), 18.

9. Interview with Jim Martz.

10. Professor Valeriano Ruiz Herandez, quoted in R. Hansen and J. Martin, *Photovoltaics for Rural Electrification in the Dominican Republic* (Lowell, MA: University of Lowell Photovoltaics Program, 1987), 104.

11. Interview with Patrick Jourde.

12. Interview with Bernard McNelis.

13. Interview with Patrick Jourde.

14. Programme Territorie de la Polynesie Francaise–CEA–AFME, "Les Energies Renouvelables en Polynesie Francaise," "Le Fare Solaire." Brochure.

15. Interview with Patrick Jourde.

16. P. Jourde, "An International Programme on PV Rural Electrification in Developing Countries," *Eleventh E.C. Photovoltaic Solar Energy Conference* (Montreux, Switzerland, 12–16 October 1992) (Dordrecht: Kluwer Academic Publishers, 1993), 1608.

17. Interview with Bill Mahoney.

18. Interview with Patrick Jourde.

19. Ibid.

20. D. Eskenazi, "Evaluation of International Projects," September 1986, SAND-85-7018/2 DE 87 002943, B-23.

21. Almost all rural electrification projects on every continent have received immense subsidies. In most instances, as exemplified by the Tahitian experience, the photovoltaic option seems the least expensive.

22. Interview with Lalith Gunaratne, who has worked in the photovoltaics field in Sri Lanka since 1984.

23. Ibid.

24. "How to Sell the Sun at the Equator? Solar in Africa," *Ecotec Resources bv* (1994): 4.

25. Interview with Mark Hankins.

26. S. Karekazi, "Photovoltaics and Solar Water Heaters," in *Kenya AFREPREN RETS Research Final Project* (July 1990): 70; M. Kimani and M. Hankins, 1993, "Rural PV Lighting Systems," in D. Waubengo and M. Kimani, *Whose Technology? The Development and Dissemination of Renewable Energy Technologies in Sub-Sahara Africa* (Nairobi: KANGO Regional Wood Energy Programme for Africa, 1993), 91.

27. Interview with Mark Hankins.

28. Kimani and Hankins, "Rural PV Lighting Systems," 93.

29. M. Hankins, in *Renewable Energy in Kenya* (Nairobi: Motif Creative Arts, 1987), 107.

30. Interview with Mark Hankins.

31. Interview with Richard Acker.

32. "Solar in Africa," *Ecotec Resources bv*, 4.

33. C. Flavin and M. O'Meara, "Shining Examples," *World Watch* (May/June 1998), 30.

34. Interview with Richard Hansen.

35. Ibid.

36. R. Hansen and J. Martin, *Photovoltaics for Rural Electrification in the Dominican Republic* (Lowell, MA: University of Lowell Photovoltaics Program, 1987), 104.

37. Interview with Richard Hansen.

38. "Bella Vista Sousua, letter to Richard Hansen, August 26, 1984." (Courtesy Richard Hansen.)

39. Interview with Richard Hansen.

40. Hansen and Martin, *Photovoltaics for Rural Electrification*, 4.

41. Ibid., 84.

42. Enersol Associates, *Solar Electricity for Rural Development* (Sommerville, MA: ENERSOL Associations, 19??).

43. P. Covell and C. Morales, "Aprovechan Electricidad del Sol en Campos de Puerto Plata," *Desarrollo, Supplemento Listin Diario* (28 November 86): 116.

44. Global Transition Group Pamphlet. (Courtesy Richard Hansen.)

45. Interview with Richard Hansen.

46. Ibid.

47. Ibid.

48. Soluz, Inc., "Business Plan" (January 1997), 2–3.

49. Interview with Neville Williams.

50. Ibid.

51. Ibid. In just three years, SELCO has sold three thousand photovoltaic systems in southern India. (Interview with Neville Williams.)

52. Ibid.

53. Interview with Guy Oliver.

54. "East and Southern Africa Wind and Solar Draft Survey" (April 1994), 17.

## Chapter Thirteen
# Solarizing the Electrified

In the mid-1970s and early 1980s, when the governments of developed countries began to fund solar energy programs, they tended to favor large-scale centralized photovoltaic plants over small, autonomous individual rooftop units because that is how electricity had been produced with other power sources. "The vision was huge solar farms—gigawatts of solar cells," attests a participant in the design of these large-scale photovoltaic installations.[1]

However, starting with solar pioneer Charles Fritts, who built the first selenium module in the 1880s, many have believed that photovoltaics is better suited for point-of-service placement. When Fritts boldly predicted that his selenium solar panels might soon compete with Thomas Edison's coal-fired power plants, he had no intention of constructing large-scale generating stations. Rather, he believed, solar arrays were "intended principally for what is known as 'isolated' working, i.e., for each building to have its own plant."[2] Ninety-two years later, the Shell Oil Company, described by Lester Brown, president of the Worldwatch Institute, as one of the more prescient firms in the petroleum industry, restated Fritts' notion: "In our opinion, the dispersed generation of [photovoltaic] energy—in shopping centers, small manufacturing plants, homes, and apartment complexes—affords the earliest opportunity for photovoltaics to contribute to our

SOLAR PHOTOVOLTAIC POWER FARM
25 MEGAWATTS (ONE SQUARE MILE)

PANEL

*Governments and utilities throughout the developed world envisioned huge solar-power farms, as shown in this 1974 illustration. Because electricity had always been generated from central power stations, it was assumed that power from photovoltaics would be, too.*

[America's] growing energy needs."[3] The journal *Science* concurred, suggesting that the government should give greater attention to "on-site . . . photovoltaic devices."[4]

The debate over how to situate photovoltaics intensified after several California utilities built, with government funding, multi-megawatt photovoltaic plants in the early 1980s. That was also when Markus Real, a Swiss engineer, took matters into his own hands to demonstrate that dispersed photovoltaic units on residences was a better idea than centralized photovoltaic stations.

In the 1980s, Real formed Alpha Real, a small company which installs photovoltaic systems. The company became well-known in Real's native Switzerland after it won the world's first solar car race staged in Europe in June 1985. Almost every Swiss had seen the "Mercedes Benz powered by Alpha Real" devastate the competition.

*An illustration from a 1929 French enclyopedia depicting solar modules on a residential rooftop and in the front yard.*

Capitalizing on its name recognition at home, Alpha Real continued making milestones in photovoltaic applications. For example, the company built the world's first photovoltaic-powered tunnel lighting system high in the Alps. But as Real and his company gained experience, they discovered that solar cells do not benefit economically from centralization as conventional power plants do. In a traditional power plant, each incremental enlargement of its turbogenerator results in a threefold increase in power. Therefore, size is a major factor in the cost of generating electricity, and it encourages the building of bigger power plants. The same rule, however, does not apply to electricity generated by solar cells. No economic benefits accrue by building larger and larger photovoltaic units. If the same number of solar cells is installed in smaller clusters, the cost of photovoltaics per watt remains the same. The price of photovoltaics decreases only as the number of modules produced rises.

*Markus Real, founder of Alpha Real.*

To prove to skeptics the greater value of siting photovoltaic units on rooftops instead of installing them in large, faraway generating plants, Alpha Real initiated its revolutionary Project Megawatt. Real called the program, "The answer to [the] large multi-megawatt installations" that had gained favor throughout the world in the early 1980s. To find participants, Alpha Real took to the airwaves and the print media, announcing that the firm "is looking for 333 power-station owners. Having a rooftop exposed to the sun is the only prerequisite."[5]

Zurich homeowners responded enthusiastically. "Thousands called us and said, 'Hey, that's interesting,'" Real recalled. More than 333 units were sold, each having a capacity of three kilowatts, for a total of one megawatt of dispersed power. Project Megawatt taught the

*Publicizing its drive to solar electrify 333 Zurich homes, Alpha Real announced in this television ad that the firm "is looking for 333 power-station owners."*

developed world that "it is on the building where photovoltaics should be placed."[6]

Since then, more people have come to realize that if each building could act as its own electrical producer, it would eliminate much of the capital costs inherent in building a centralized power plant. Steve Strong, an architect and long-time advocate of residential photovoltaic systems, described the added expenses incurred in putting up a field of photovoltaic arrays for large-scale electrical generation: "You've got to buy land, do the site work on it, dig lots of holes, pour lots of concrete, dig trenches, bury conduits, build foundations and support structures, buy a huge inverter to change the photovoltaic-generated DC current into AC, construct a building in which the inverter is placed, purchase switch gear, and a switch yard and transformers, and, because your station is usually far away from where people live, you have to spend money on transmission lines to get the electricity where the need is. You have spent a great deal of money and you have yet to buy a single solar cell!"[7]

None of these outlays in time, money, or effort are necessary if the solar modules are placed on the buildings where the electricity will be used. "It makes sense, absolute sense," argued Real. "The roof is there. The roof is free. The electrical connections are there."[8] In fact, in densely populated countries, such as those in Europe and Japan, the high cost of land rules out the use of photovoltaics unless it is placed on buildings. Switzerland, for example, "cannot afford to waste large tracts of land for the sole benefit of photovoltaic plants."[9] In Germany, too, "the prices for real estate are particularly high [and] to power a standard TV set for . . . 3 hours [requires] a space of 2 square meters [22 square feet] of solar cells," according to a German photovoltaics specialist. Hence, "it [is] quite logical to consider . . . buildings as space for photovoltaic cells."[10]

Donald Osborn, director of alternative energy programs at the Sacramento Municipal Utility District (SMUD), outlined other advantages of on-site photovoltaic electrical generation for both the consumer and the utilities. "You reduce the electricity lost through long-distance transmission," Osborn stated, which runs to about 30 percent on the best-maintained lines. Structures with their own photovoltaic plants decrease the flow of electricity through distribution lines at substation transformers, "thereby extending the transformers' lives." "And for a summer daytime peaking utility," Osborn added, "you can offset the load on these systems when the demand for electricity would be greatest," helping to eliminate "brownouts in the summer and early fall." On-site photovoltaic-generated electricity also makes renewable energy economically more attractive than power generated by a large solar electric plant because it "competes at the retail level rather than at the wholesale level" with other producers of electricity.[11] Large-scale solar plants have to generate electricity at 2¢ to 2.5¢ per kilowatt hour to be competitive with other commercial systems, while residential photovoltaic units can make electricity at 10¢ or more per kilowatt hour and still be economically viable because that is the price homeowners would have to pay anyway.

Putting photovoltaics on buildings where electrical networks already exist eliminates the cost of storage. In most of the developed world, such micropower systems would not only produce enough electricity for in-house use, but they also would generate excess power that could be sent back to the utility during the day. At night or in inclement weather, utility-generated electricity would return the needed power.

Just as Real was finishing his rooftop program, chance had it that S. David Freeman, then director of the Sacramento Municipal Utility District, came to Switzerland to spend a few days at a chalet owned by a close friend of Real's. Freeman had gained fame by shepherding the closure of the Rancho Seco nuclear power plant, which made SMUD the first utility in the world to pull the plug on an operating "nuke." Freeman became an

*A field of photovoltaic panels next to the Rancho Seco nuclear power plant.*

even greater figure among environmentalists when, during his tenure, a two-megawatt photovoltaic plant was installed adjacent to the huge cooling towers that had once served Rancho Seco. At the chalet, Freeman talked of plans to build larger photovoltaic plants, arguing that economies of scale would make the technology more affordable. Real chimed in that if Freeman were talking about turbines, he would be correct, but he had it wrong when it came to photovoltaics. "For one megawatt of photovoltaics you need ten thousand square meters [108,000 square feet] of panels," Real told Freeman. "Whether you put ten thousand square meters on one spot or disperse one hundred square meter panels on one hundred roofs is irrelevant in terms of cost of the photovoltaic material."[12] He then went on to tell about Alpha Real's successful rooftop program.

Real had expected that his revolutionary experiment and its far-reaching implications would turn Freeman's thinking around and fire him up to repeat in California what the Swiss engineer had accomplished in Switzerland. "I anticipated that David Freeman would say, 'Wow! Yes, that's right,'" Real recalled. "But from Freeman's comments and demeanor, he didn't even seem interested. I felt at the time I had never talked to anyone who didn't understand me as much as when I spoke with [him]." So Real got the surprise of his life when, following Freeman's return to the States, he received a call from his friend David Collier, who worked closely with Freeman. "What did you say to Dave Freeman?" Collier asked. "He told me we should start doing residential photovoltaic installations. Begin with a hundred and then do more every year."[13]

So began Sacramento's innovative Photovoltaic Pioneer Program, where the utility installs four-kilowatt power plants on the roofs of volunteering rate payers. To date, four hundred thirty-six rooftops have been fitted with photovoltaic modules, for a combined capacity of almost two megawatts. The utility plans to similarly equip one thousand more residences in the near future, increasing substantially the number of "low-cost power plant sites." The program in Sacramento substantiates what Real had been saying all along and has made believers of the utility executives at both the Sacramento Municipal Utility District and the San Diego Gas and Electric Company. Skip Fralick of San Diego Gas and Electric called rooftops "'free land' . . . [since] it needs no development, environmental impact statements, or extensions of transmission lines."[14] Roof installations, Osborn concurs, have allowed his utility to site "photovoltaic power plants all across Sacramento with little trouble or expense."[15]

*After conversations with Markus Real, S. David Freeman, former director of the Sacramento Municipal Utility District, changed the focus of the utility's photovoltaic program from large-scale power generation to rooftop system. SMUD's large purchases have helped to bring down the price of photovoltaics considerably, with the expectation that by 2002 solar electricity will cost the same or less than utility-generated electricity.*

Large purchases of photovoltaic modules have also helped to lower their price, just as Real had predicted. Costwise, "we're moving right along," according to Collier, who now manages SMUD's photovoltaics program. "This year we installed systems at $5.30 a watt [generating electricity at between 16¢ and 18¢ per kilowatt hour]. We expect to be below $5 a watt next year. And in 2002, we have firm commitments that we will be under $3, which makes this stuff really competitive with commercial power. We're looking at generating electricity at less than 10¢ per kilowatt hour."[16]

Not only has the utility's active role in purchasing photovoltaics helped to "shave several years off the commercialization path for the technology" in the wired market, it has benefited the utility as well. Those on Wall Street "constantly point to our renewable energy programs when upgrading our bonds," said Don Osborn. "Photovoltaics especially is viewed by many bond-rating companies as an indication of the forward-looking, flexible nature of a utility's management and its ability to predict and cope with change."[17] As photovoltaic systems accumulate on the rooftops of the Sacramento area, the utility has also gained a wealth of experience, which, in the upcoming deregulated market, could prove a windfall. According to David Collier, his "phone rings off the hook with people [from] outside the district wanting to figure out how they could buy some of its clean energy."[18]

Other American utilities have followed SMUD's lead. Edison Technology Solutions plans to install photovoltaics on one hundred elementary and intermediate school roofs in the Los Angeles Basin in 1999. S. David

Freeman, now at the helm of the Los Angeles Department of Water and Power, has decided to continue the photovoltaics work he began in Sacramento. At least twenty-five hundred Los Angeles rate payers will generate their own power with solar cells over the next five years.

GPU Solar, a subsidiary of GPU, the American east coast utility that ran the Three Mile Island nuclear power plant, and Green Mountain Energy, an energy marketer, have begun to aggressively sell photovoltaic systems to homeowners. They hope to make buying them "as simple as buying a fridge," asserts Green Mountain's John Quinney.[19] These new companies will handle all the daunting paperwork, such as arranging interconnection agreements with the local utility. They will also obtain the available rebates, warranties, and financing and will oversee the installation.

Alpha Real's Project Megawatt sparked a revolution in the use of photovoltaics in other parts of the developed world as well. Japan, for example, has embarked on an even more ambitious rooftop program than its American counterparts. Starting in 1994 with seven hundred residential installations, the Japanese expect to have one hundred eighty-five megawatts of photovoltaic power generated on sixty-two thousand roofs by 2001.[20] The Japanese government supports the program because it wants to significantly increase the production of photovoltaics with the intent of drastically lowering the price of solar cells. But the Germans have outdone everyone. A one thousand-rooftop program initiated in the early 1990s has evolved into a grand scheme of putting up one hundred thousand one- to five-kilowatt units on as many rooftops between 1999 and 2005. By that year, the Germans hope to have installed three hundred megawatts of photovoltaic material, more than twice the world production in 1998![21] The government will provide low-interest loans because it believes that "photovoltaics will soon become competitive, so [it] had better get some experience with the new technology."[22] That so many rooftop programs are currently going on, Real believes, proves that "the concept is now commonly understood and accepted. There is no doubt that a large movement toward the rooftop approach is in progress."[23] "Nobody is even talking about big photovoltaic plants, central power stations anymore," confides an executive of Kyocera, the largest producer of solar cells in Japan.[24]

Project Megawatt also helped to revolutionize the buying and selling of electricity between mini-producers and utilities. When participants in Project Megawatt produced more electricity than they needed, they sold it

*An apartment complex in Bremen, in northwest Germany, a participant in that nation's one hundred thousand-roof photovoltaic program.*

to the local utility for 2¢ per kilowatt hour. But when they needed electricity—for example, at night—the local utility charged them 12¢ per kilowatt hour. Real attributed the unbalanced billing procedure to the fact that "the idea that electricity doesn't flow solely from the central power station to the consumer, but had become a two-way street with each consumer also becoming a producer, was, of course, something new for utilities to chew on." But the unfairness of such pricing infuriated Alpha Real's clients, whom Real described as people of influence: "They were not 'green' or left, but doctors and lawyers." And when these professionals protested, Real recalled, "There was so much pressure from the public sector" that the electric company relented.[25] It agreed to buy from and sell to the independent electricity producers at the same price. Net metering, the establishment of equitable rates when large and small electrical producers interact, has become the accepted way of doing business in Germany, Japan, Switzerland, and twenty-three U.S. states. It makes photovoltaic household systems more economical in the developed world by reducing the payback time from around forty-five years to twenty.[26] It also gives a psychological boost to homeowners who install solar cells on their rooftops. "The idea of being able to spin a utility meter backward [that is, sell electricity back to the utility] really appeals to people," observed one photovoltaics engineer.[27]

With the inclusion of photovoltaics in mortages managed by two of America's largest lenders, Fannie Mae and GMAC Inc., photovoltaics for new houses becomes economically appealing. Amortizing a photovoltaic installation over a thirty-year period makes purchasing solar electricity no different than paying the electric bill each month—except that the power plant belongs to the homeowner and it doesn't pollute.

To make photovoltaics as financially appealing as possible, Markus Real, among others, "realized [that] installations had to be better integrated into the construction and that components had to be developed which builders could put in routinely."[28] Joachim Benemann, a German photo-voltaics engineer, agreed with Real. As he saw it, the main problem with rooftop systems was that the modules were add-ons. They were placed on racks, which resulted in a double roof or double facade and a lot of rewir-ing. To eliminate the duplication and the extra work, Benemann suggested that photovoltaics be integrated into the facades or rooftops of buildings.[29]

Benemann originally worked for Flachglas, which is now owned by Pilkington, one of the oldest and largest glass companies in the world. Flachglas had entered the solar field by making reflectors for the Luz solar thermal plant in Daggett, California. In its heyday, the plant added eighty megawatts of electrical generating capacity each year, when only twenty or thirty megawatts of photovoltaics were being produced annually world-wide. The photovoltaics business "looked like peanuts" compared to the solar reflector business, Benemann recalled. "So we concentrated on the big business, which was solar thermal. But when we saw the demand for solar reflectors collapse, then we said, 'O.K., photovoltaics might become an interesting business for us.'"[30]

Flachglas, however, felt it could not go head-to-head with the large electric and oil companies that dominated the photovoltaics industry. "We asked ourselves," Benemann recalled, "where is there a niche for us? What kind of application are we better at than all the rest?"[31] It quickly became clear that the company's success depended on entering the architectural market where Flachglas had the experience and the contacts that the other photovoltaics companies, including Markus Real's, lacked.

Benemann's company came up with a general concept for packaging architectural photovoltaics. "Our idea was [that] if glass is becoming such a very advanced and innovative aesthetic building material, which has a lot of special features, such as sun reflecting, heat insulation, and sound- and bulletproofing, why shouldn't the glass used for the skin of the building also produce electricity? So we said, let's encapsulate the solar cells into the glass and let's do it in a very pleasing aesthetic and optical way so we can meet the architect's needs and increase interest for this new technology."[32]

Flachglas found that a proprietary transparent resin, which it used to adhere plates of glass together for soundproofing air terminals and adjacent hotels, could also be used to embed solar cells into architectural glass. Working with a proven technology assured architects that the new "photovoltaic glass" was not a prototype but a commercial product. Like traditional architectural glass, building-integrated photovoltaic glass can be customized for almost any job. It offers great flexibility in shape and size, the largest pane being sixty-three square feet (six square meters), and provides thermal insulation and protection from sound and weather.

Using photovoltaics as a building skin completely changes the way of calculating the economics of photovoltaic installations: The old method, considering electrical generating costs, no longer holds. Instead, the expense of photovoltaic building material must be compared with the price of other facade coverings. Marble and granite, for example, cost more, and neither generates one watt of electricity. A photovoltaic building skin can produce up to one-third the energy used in an office building, or about enough to light up the entire interior.[33] Purchasing a building skin that also generates electricity at an equivalent or lower price than conventional materials is good business. This is especially true in densely developed urban sites where most commercial buildings are built and where "power tends to cost a lot more due to the high cost of adding new capacity," Gregory Kiss, a Manhattan-based architect, explained. "Here the value of photovoltaics becomes quite high."[34] There is also a good match between the availability of solar energy and the power needs of office buildings. Most of a building's electrical consumption occurs between 9AM and 5PM when both employees and the sun are usually at work. And since the solar cells can be positioned in the glass as close together or as far apart as desired, their configuration can block out or let in as much sun as the architect wishes, helping to save on illumination, air conditioning and heating devices, and the energy they consume.

Factors other than economics also enter into building skin selection. Aesthetics and the image the building owner wishes to convey often override financial concerns. Banks and insurance companies, for example, have

*Left: Photovoltaic shades can help control the amount of solar heat entering a building, thus reducing the electricity consumed by air conditioning.*

*Right: Flachglas' first installation that integrated solar cells into a building's facade. Here photovoltaics replaced the vertical glass on the south side of the structure.*

*Left: Solar cells integrate well into buildings where large amounts of glass are used as a covering.*

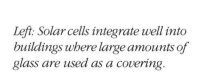

traditionally favored imposing facades that connote strength and stability. With many companies realizing that there is an economic benefit in being perceived as environmentally responsive, a photovoltaic facade boldly displays concern for the welfare of the planet.

The first fully building-integrated photovoltaic facade was built in 1991; it covered the south side of a utility company's administrative offices in the small town of Aachen, Germany. To the utility's surprise, its pioneering building became a mecca for architects and engineers. The company had to hire two additional engineers to answer the inquiries of the many visitors who flocked to the site.

Following the success of the Aachen project, under Benemann's leadership, Flachglas, and its parent company Pilkington, has aggressively championed building-integrated photovoltaic material. Knowing that unfamiliarity presents the greatest barrier to the success of a novel technology, especially in the construction business, with its many disparate professions, Pilkington offers wary architects and investors a complete service. Prospective buyers who hesitate because they have never worked with photovoltaics are told,

"Please! Don't worry. We'll do everything, including the design, the engineering [Pilkington also owns one of the foremost electronics companies in Europe], and we'll provide assistance in financing and insurance." To make sure everyone feels comfortable, Benemann added, Pilkington "will provide a maintenance contract, just like elevator companies do, and it will check and repair the system."[35]

Pilkington's farsighted marketing approach appears to be working. The company has completed more than one hundred large building-integrated photovoltaics projects and has more in the works. Europe's largest railway terminal—Berlin's new central station, which is presently

*These solar modules also shade the interior from the excessive heat of the summer sun.*

*The first major American commercial building to use photovoltaics as part of its facade—4 Times Square—is now under construction in the center of Manhattan. The view is looking northwest along 42nd Street.*

under construction—will have a photovoltaic glass roof over its passenger platforms. The home of the Academy of Further Education in Herne, Germany, will soon boast the largest amount of photovoltaic glass on a single structure in the world, one hundred thousand square feet (9,290 square meters) of south/southwest wall and roof, with a total output of one megawatt. Following the growing trend in Germany to incorporate photovoltaic material into structures, the government of North Rhine–Westphalia, Germany's most populous state, now requires all new public buildings to be built with photovoltaic material.

The popularity of building-integrated photovoltaics in Germany stems from the increasing desire for glass facades on large structures, the country's powerful "green" constituency, and the beneficial image visible photovoltaics provides. Government and utility assistance with photovoltaics projects, such as buying back excess solar-produced electricity at premium rates, reflects the power environmentalists wield in German politics.

Across the Atlantic, the Durst Organization is now constructing the first major office building being built in New York City in the 1990s. The south and east sides of the 38th to the 45th floors of the 48-story skyscraper, where the most sunlight will strike, will have a photovoltaic skin. 4 Times Square, on the corner of Broadway and 42nd Street, "will be the opposite extreme of the traditional remote photovoltaic installation," said Gregory Kiss, the architect responsible for the photovoltaic facade. "Here we are planting the photovoltaics flag in Times Square, the crossroads of the commercial world. 4 Times Square leads the way to large-scale generation of clean, silent solar electricity at the point of greatest use, in urban centers like New York City."[36] Indeed, *Architectural Record*, the influential journal of the profession, believes that the use of photovoltaic facades has helped "solar energy . . . break out of its experimental corral . . . and become a cost-effective supplemental energy option that architects should routinely consider."[37]

## Notes & Comments

1. Interview with Richard Swanson. The major utilities of the developed world strongly influenced their governments' push for large-scale photovoltaic plants. The Electric Power Research Institute, for example, wrote in 1983, "for solar cells to make a significant impact on electricity production in this country . . . [t]hey will have to go for the real thing: bulk electric power generation." "Photovoltaics: The Excitement Is Still There," *EPRI Journal* (July/August 1983): 2.

2. C. E. Fritts, "Note," in W. Siemens, "On the Electromotive Action of Illuminated Selenium, Discovered by Mr. Fritts, of New York," *Van Nostrand's Engineering Magazine* 32 (1885): 515.

3. Shell Reports, *Solar Energy*, rev. (Houston, TX: Shell Oil Company, 1978).

4. A. Hammond and W. Metz, "Solar Energy Research: Making Solar after the Nuclear Model?" *Science* 197 (15 July 1977): 243.

5. "Alpha Real sucht 333 Kraftwerbesitzer, MEGAWATT Solarkraftwere für unsere Umwelt." Company brochure.

6. Interview with Markus Real.

7. Interview with Steve Strong.

8. Interview with Markus Real.

9. J. Gay et al., "Architectural PV Integration at the EPFL," *Eleventh E.C. Photovoltaic Solar Energy Conference* (Montreaux, Switzerland, 12–16 October 1992) (Dordrecht: Kluwer Academic Publishers, 1993), 668.

10. J. Benemann, "Photovoltaics in Buildings—A Step Towards the Future." (Courtesy Joachim Benemann.)

11. Interview with Donald Osborn.

12. Interview with Markus Real.

13. Ibid. David Collier confirms the conversation he had with Markus Real, and credits the discussions Real had with Freeman in Switzerland for changing the direction of the photovoltaics program at the Sacramento utility from centralized installations to individual rooftops.

14. D. Osborn, "Commercialization of Utility PV Distributed Power Systems," *Solar 97* (Washington, D.C., American Solar Energy Society, 1997), reprint.

15. Interview with Donald Osborn.

16. Interview with David Collier.

17. Interview with Donald Osborn.

18. Interview with David Collier.

19. Interview with John Quinney. Before companies like GPU Solar and Green Mountain Energy offered to facilitate the installation of photovoltaic systems, homeowners in many states could legally install photovoltaics and connect them to existing power lines. However, local utilities usually made the procedure so troublesome that most homeowners who wanted to produce their own electricity eventually gave up on the idea.

20. S. Strong, "An Overview of Worldwide Development Activity in Building-Integrated Photovoltaics," *Proceedings of the 1st International Solar Electric Buildings Conference*, vol. 1 (Boston, Massachusetts, 4-6 March 1996) (Greenfield, MA: Northeast Solar Energy Association, 1996), 18.

21. Th. Erge et al., "PV on Residential Buildings: Examples and Results of the German 1000 Roofs Programme," *Proceedings of the 1st International Solar Electric Buildings Conference*, vol. 1 (Boston, Massachusetts, 4-6 March 1996) (Greenfield, MA: Northeast Solar Energy Association, 1996), 22.

22. Interview with Dr. Wolfgang Palz.

23. Interview with Markus Real.

24. Interview with Al Panton, manager of Kyocera's U.S. solar operations.

25. Markus Real, 1994, video interview. (Courtesy Mark Fitzgerald.)

26. "Solar Power: Consumers Cash in," *Popular Science* 257 (January 1998): 32.

27. "Power to the People," *Scientific American* 276 (May 1977): 44.

28. Interview with Markus Real.

29. J. Benemann, "Photovoltaics in Buildings as a Step towards the Future." (Courtesy Joachim Benemann.)

30. Interview with Joachim Benemann.

31. Ibid.

32. Ibid.

33. Optisol, "Solar Energy Facades." Pilkington International brochure.

34. Interview with Gregory Kiss.

35. Interview with Joachim Benemann.

36. Interview with Gregory Kiss.

37. "Product Reports 1996. Record's Panel Picks Top Products," *Architectural Record* 184 (December 1996): 64.

Chapter Fourteen
# Better Cells, Cheaper Cells

For several decades, solar cells have been the least expensive and most reliable power source for small-scale electrical devices located away from utility lines. When photovoltaics produced electricity at $144 per kilowatt hour, its use was confined to the most remote of all applications—space—where the choice of primary long-lived power sources was extremely narrow. In the late 1970s and early 1980s, as the price of photovoltaic electricity dropped to less than $1 per kilowatt hour, solar cells began to replace the earlier-used, but now more expensive—and always less dependable—remote power sources such as generators and batteries. In the last decade, with the price of solar electricity falling below 25¢ per kilowatt hour, the technology has become an even more attractive choice for remote small power users.[1] In fact, its price decrease accounts for photovoltaics' entrance into the suburban and urban landscapes of the developed world.

Emergency call boxes along California's roadways were one of the first visible applications of photovoltaics in the state's highly populated areas. By 1993, a single manufacturer had installed over 11,250 such systems. The savings were immediate. The city of Anaheim, California, for instance, would have had to spend almost $11 million to connect its 1,100 call boxes to utility lines. Using photovoltaics as the power source, the city

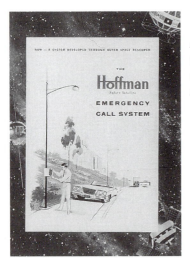

*Left: In 1961, solar visionary Leslie Hoffman proposed the use of solar cells to run emergency call boxes along freeways in Southern California. Below: Twenty-eight years later, solar call boxes have become common throughout the state.*

*Solar panels provide the electricity for lighting bus shelters along the Strip in Las Vegas.*

*Flashing lights run by solar cells warn motorists to slow down for school children.*

paid only $4.5 million. When Carrollton, Texas, a suburb of Dallas, considered placing warning lights on speed limit signs near schools, planning engineers calculated that underground trenching from the nearest commercial power source to all the signs would cost over $200,000. Installing solar modules to run flashing lights cost half that.[2] When Sacramento needed to put in new streetlights, it was discovered that ripping up the streets to lay wires would cost more than $3,000 per light. A photovoltaic system cost $500 less, and there was no disruption to the roadways.[3] When faced with spending tens of millions of dollars to tear up streets and yards to add underground power lines in the affluent community of South Pasadena, Southern California Edison, now Edison International, also chose photovoltaics. The utility strategically placed photovoltaic panels around town for under a million dollars. As Steve Taylor, who worked on the project explains, "It's just a typical application of using distributive generation versus centralized generation. We actually took the production of electricity out to the customers."[4] "Cost effectiveness for photovoltaics comes not from the energy saved in applications such as these," observed John Thornton, a photovoltaics specialist, "but from not having to waste time and money digging up an existing street and then paying the utility for an electrical connection." "Of course," Thornton added, "once [the photovoltaic system] is in, the electricity is free."[5]

Building-integrated photovoltaics has also brought the use of solar cells into the developed urban world. But to realize the dream of solar cells covering the world's rooftops on a large scale, their price must drop from the current $5 per peak watt to $3 or less, producing electricity at 10¢

*No wires—underground or overground —were needed to light this parking area. Each streetlamp has its own solar module providing its electricity.*

to 15¢ per kilowatt hour or less. Many believe that this reduction will come only with changes in cell manufacture.

Presently crystalline silicon remains the dominant photovoltaic material, basically the same substance that was discovered at Bell Laboratories in the early 1950s. Its continued widespread use rests on the fact that, as one expert explained, "It works and it works for a long time."[6] As Darryl Chapin had predicted in 1956, there has been "a [significant] reduction in the cost of purification and fabrication of finished [silicon] cells." However, despite the many advances made in the production of crystalline silicon solar cells over the last forty or so years, the manufacturing process begins, just as in Chapin's day, "with high purity silicon." Though its price has come down considerably, it remains, as Chapin pointed out in the 1950s, "not a cheap material."[7] In fact, the crystals for single-crystal silicon cells continue to be grown as they were in Chapin's time: by melting silicon at 2570°F (1410°C) in a rotating vat and then drawing it up and out of the melt by a puller spinning in the opposite direction. This produces a single-crystalline cylinder 7 to 8 feet (2 to 2.5 meters) tall.

Attempting to bring down the price of solar cells, Chapin tried to find new ways to make crystalline silicon. For instance, he tested cells fabricated from polycrystalline silicon, that is, crystalline silicon cast into ingots. Casting silicon is less complex than growing a single crystal, and the result is smaller multiple crystals rather than one large one. But, the activated charges in the polycrystalline material had great difficulty crossing from

*The procedure for preparing single-crystalline silicon solar cells. Material loss due to cutting is high: 16.4 grams of silicon are required for one peak watt.*

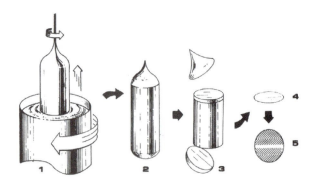

(1) SILICON IS MELTED AT 2570°F (1410°C) IN A ROTATING VAT. AS IT BEGINS TO CRYSTALLIZE, IT IS DRAWN UPWARDS FROM THE VAT BY A COUNTER-ROTATING ASSEMBLY. THE RESULTING SOLID CYLINDER (2) IS ROUNDED AND THE TOP AND BOTTOM ARE CUT OFF (3). THE CYLINDER IS SLICED INTO WAFERS (4) WHICH ARE PROCESSED INTO SOLAR CELLS (5).

crystal to crystal, finding themselves trapped in the crevices or boundaries between them. Hence, relatively few charges made it through this obstacle course to the metal contacts. Chapin, therefore, obtained dismal results—an efficiency of only 1 percent.[8] Over the next two decades, experimenters increased the efficiency of polycrystalline silicon cells by a few percentage points, but never beyond the 6 percent mark. The failure to reach higher efficiencies lead experts to conclude that good solar cells could not be made with this material.

"This was disproved, though, by Joe Lindmayer," commented Fritz Wald, who has been involved in solar cells for many years, and simultaneously by Wacker Chemitronics in Germany.[9] Lindmayer was a founder of Solarex, which followed Solar Power Corporation in pioneering the manufacture of photovoltaics solely for earth applications. In the early days, the supply of single-crystal material "was like a roller coaster," according to a solar expert who worked closely with Lindmayer. "The supply went up and down depending upon the needs of the semiconductor industry. Sometimes it was abundant and sometimes very scarce." To guarantee a steady supply, the company could have invested in its own crystal-pulling devices, but they were very expensive. "We just didn't have the money for it," the expert admitted. "So Joe experimented with a number of other options. One of them was to cast the silicon."[10]

Dr. Lindmayer was regarded as one of the best technical people in the solar cell business during the 1970s: If anyone could make polycrystalline silicon work, he could. He found that past efforts had failed because the crystals produced were too small, and, therefore, the area of the boundaries was very large. So he set out to grow relatively large crystallites. In the process, Lindmayer discovered that crystal size depended on the speed at which the ingot cooled: Fast cooling meant small crystals. By slowing down the cooling process, Lindmayer came up with the size needed to make an efficient solar cell, approximating the performance of single-crystalline silicon.[11] Though he had solved the supply problems that Solarex had experienced, the development of a good polycrystalline solar cell did not bring about the expected price reduction for photovoltaic modules.[12] For one thing, the casting procedure still begins with expensive highly refined sili-

*A large polycrystalline silicon ingot. Ingots such as this are sawn into thousands of very thin wafers, which are made into solar cells.*

con, and the cast blocks, which weigh 120 pounds each, still must be sliced into extremely thin wafers to make solar cells. Single-crystalline cylinders also have to be reduced to segments three hundred to four hundred microns thick (one to two one-hundreths of an inch). Cutting such thin pieces from very large starting materials creates a lot of waste. "You eventually make as much sawdust as you do cells," Wald observed.[13] Half of the very expensive silicon ends up as trash.

To put an end to the terribly wasteful sawing phase, a number of people have come up with the idea of growing silicon crystals in sheets, as thin as wafers, directly from molten silicon. However, only Mobil Solar Energy Corporation, recently purchased by the German photovoltaics firm ASE, has fully developed and commercialized sheet silicon.

What became known as Mobil Solar Energy Corporation began as the vision of one man, Dr. Abraham Mlavsky, who ran a small research and development company called Tyco. Mlavsky had the idea of producing a continuous thin ribbon of silicon that would be separated into suitable lengths, processed into solar cells, and placed directly into modules. This was no harebrained scheme. In the 1960s, Mlavsky and his personal technician, Harry Labelle, had successfully drawn continuous shapes of sapphire crystal from molten sapphire. Mlavsky then decided to substitute a silicon melt and apply this technology to the photovoltaics industry.

The process of making silicon ribbon begins with a vat of molten silicon, much like the method for single-crystalline silicon cylinder production. This vat, however, remains stationary and a graphite die is suspended into it. The molten silicon rises naturally into the die, which shapes it into a thin sheet as it passes through a narrow slot at the bottom of the die. Crystallization is induced right above the die, as the very hot liquid silicon

*The method for producing crystalline silicon ribbon.*

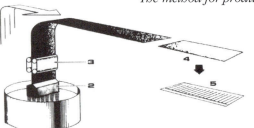

(1) SILICON IS MELTED AT 2570°F (1410°C) IN A VAT. AS IT CRYSTALLIZES, IT IS DRAWN OUT OF THE VAT THROUGH A GRAPHITE DIE (2) BY A BELT PULLER (3). THE RIBBON OF SILICON, WHICH IS MUCH THINNER THAN THE SINGLE CRYSTAL WAFER, IS CUT INTO CELL BLANKS (4) WHICH ARE PROCESSED INTO SOLAR CELLS (5).

immediately freezes into a ribbon as it is mechanically pulled out into the much colder air.

In 1971, NASA, seeking lighter-weight solar cells for spacecraft, gave Mlavsky the chance to produce solar cells from ribbon silicon. However, the crystalline silicon he produced did not measure up to the conversion efficiencies of crystalline silicon solar cells fabricated from wafers, so NASA temporarily lost interest.

Through its subsidiary, Jet Propulsion Laboratory (JPL), NASA renewed funding for the ribbon silicon process when it shifted its attention to terrestrial uses of photovoltaics in the autumn of 1973. JPL judged the process as "the key . . . to low-cost silicon," which the agency believed would someday lead to its use for mass-produced electricity.[14] Mobil Corporation shared the space agency's faith in Tyco's technology, pouring millions of dollars into the company.

Experts believed that the greatest barrier to increasing efficiency was eliminating the contaminants that found their way into the ribbon silicon during the shaping process. "Whoever solves [this]," one consultant predicted, "really has something going."[15] The infusion of new capital allowed Tyco to discover that impurities in the graphite die were the culprit. When the die's graphite was purified, cell efficiency rose to a respectable 10 percent.[16]

More difficulties appeared, however, despite the experts' expectations. In late 1978, after building and operating seven ribbon machines, scientists learned that "it was very hard to realize Mlavsky's vision" of the high growth rates of the ribbons necessary for "the very high volume, low cost production required for electric utility use of photovoltaic devices."[17] Much to their dismay, researchers at what was now known as Mobil-Tyco found that when the pulling speed was increased too greatly, in attempts to meet production goals, the ribbon buckled and the quality of the silicon crystals dropped precipitously. This lowered cell efficiency to unacceptable levels.

Many working on the project argued that building and running more ribbon machines would produce the desired quantity in an acceptable amount of time without overtaxing the material. A review of the numbers, however, showed that constructing the hundreds of machines necessary for the large output desired did not make sense economically. For reasons of space alone, the idea had to be dismissed.

By 1981, Mobil-Tyco had chosen instead to develop one machine that would produce many ribbons simultaneously. But this did not work out either. "The problem with the multiple ribbon machines was they became too complicated," Wald recalled. "They had too many mechanical parts and control elements, one of which would always fail."[18]

Ribbon growth in single or multiple machines presented another seemingly insurmountable obstacle: The crystallization at the ribbon's edges proved impossible to control in a precise way. One smart scientist suggested that the company change the die to shape tubular polygons, because a polygon would have no exposed edges and yet maintain the flatness necessary for solar cells. Wald believes "this was probably the most important development in sheet technology."[19] The polygon's greater diameter also permits the growth of more material without increased pulling speeds. So "even though the rate is not accelerated, the width and therefore the amount of photovoltaic product [produced] is multiplied significantly," according to photovoltaics analyst Bob Johnson.[20]

By 1988, the company had committed itself to fabricating sheets of silicon from eight-sided polygons. Now engineers and scientists had to build new machines. Having decided that the octagons were to be five meters high (16.5 feet), the company had to construct a factory with a ceiling that could accommodate their height. Next, a monorail was built to transport the finished octagons to an automated laser cutting station, where a specifically designed autofocus device guaranteed highly reliable, low damage cuts for each square of the polygon. These efforts proved worthwhile. The process has cut the cost of fabricating crystalline silicon solar cells in half. The savings come from greater automation, from using far less silicon, and from the packaging efficiencies realized by the square shape of the wafer, which can fill an entire square-shaped module.

In the early 1990s, the company developed a pilot plant that could produce twenty-five megawatts of solar cells, if demand exploded. It had planned to sell large quantities of cells to utilities in the high desert regions of the western United States to help power residential air-conditioning units. Mobil believed that photo-

*An octagon of crystalline silicon and solar cells cut from an octaganal tube.*

voltaics could help electric companies meet peak demand, which usually occurs in the early afternoon during late summer when both air conditioning use and electrical production by the solar cells are at their maximum. According to company strategists, without photovoltaics the utilities would have had to purchase additional power from other utilities at premium prices. However, those strategists had not anticipated the falling price of natural gas. This gave utilities a cheaper alternative for generating excess power, and it checkmated the company's marketing plans.

Forecasts predicted that the price of natural gas would remain low for quite some time. Not seeing any other large markets for photovoltaics, Mobil decided to get out of the solar cell business in the mid-1990s. ASE Americas, a subsidiary of a large German holding company that also makes and sells photovoltaics, felt otherwise, and it bought the company. Combining its expertise in photovoltaics with the cost-cutting octagon technology, ASE Americas is able to sell modules at 20 percent less than traditional crystalline silicon modules. The company produced four megawatts of solar cells in 1998 and hopes to double production every year.

During the debate over the best way to grow sheet silicon, Dr. Emanuel Sachs, then an engineer at Mobil's photovoltaics plant, proposed and patented a way to perfect the ribbon-growing process. Searching for a method that guaranteed the stability that had eluded the other attempts to make continuous ribbons, Sachs found that eliminating the die and, instead, drawing two parallel strings through the vat did the trick. A thin sheet of molten silicon spans the strings as they move through the vat, adhering to them in the same way that soap bubble solution clings to a bubble wand. As a machine pulls the strings and the attached membrane of liquid silicon out of the vat, the silicon freezes into a solid ribbon.

Sachs' process took Mlavsky's original idea from dream into reality. However, with the commitment to polygons, no one showed the slightest interest in Sachs' invention. Its revival came when several key employees left after Mobil sold its photovoltaics operation and formed their own company, Evergreen Solar. It remains to be seen whether the new company's process can grow enough silicon ribbon fast enough to make it competitive with, or even cheaper than, other photovoltaic technologies.

Another way to lower the price of solar cells is to direct more light onto them than they would ordinarily receive. Built of inexpensive plastic, a concentrator works like a magnifying glass. By focusing twenty times the

*A concentrator system.*

sun's energy onto a photovoltaic panel, a consumer could produce the same amount of electricity with 95% less photovoltaic material than previously necessary. Low production volume remains the only obstacle to realizing this technology's potential.[21]

Yet another avenue for low-cost cell manufacture involves depositing a small amount of a photovoltaically active substance onto an inexpensive supporting material, such as glass or plastic. Solar cells made in this fashion are called thin films. It has long been believed that this process would consume much less of the expensive photovoltaic ingredients and that it would also lend itself to automation. In the mid-1950s, while at Bell Laboratories, Darryl Chapin supported this view, hoping that fellow technologists would devise "a new process whereby a really large surface could be sensitized [photovoltaically] without the preparation of [large] single crystals."[22]

A decade later, Dr. Fred Shurland, a scientist at Clevite Corporation in Cleveland, thought he had found a way to make a cheap photovoltaic device in exactly the way Chapin imagined. Shurland substituted cadmium sulfide, a substance which also produces the photovoltaic effect, for silicon. It appeared easy to evaporate or spray a thin film of cadmium sulfide onto plastic or glass. Shurland assured his peers, "There is nothing apparent in the materials required or in the cell design or the fabrication methods to indicate that [cadmium sulfide cells] could not be made in real mass production at very low cost."[23] In 1967, an affordable solar cell ready for mainstream use seemed so near that the April 1 issue of *Chemical Week* announced that Shurland's firm "has developed a new . . . solar cell that can be mass produced at low cost, by . . . deposition of [cadmium sulfide] onto . . . plastic."[24]

Despite such sunny expectations, a usable low-cost cadmium sulfide cell has never materialized. Some Clevite cells worked well outdoors for seven or eight years; the majority, however, lost all power when exposed to the elements. Why a few lasted indefinitely and the rest fell apart remains a mystery to this day. Despite the fiasco at Clevite and a warning from

NASA that cadmium sulfide cells "are very stable [only] if you keep them from seeing light, oxygen and water vapor,"[25] the federal government and private investors, such as Shell Oil, put their money and their hopes on cadmium sulfide to mainstream photovoltaics. By the early 1980s, however, most involved in researching the material had quit because they could not come up with a long-lasting, efficient cell.

Discovering the solar device that would power America's lights and appliances with the free rays of the sun continued to lure researchers. While most laboratories interested in finding a low-cost photovoltaic device concentrated on cadmium sulfide during the 1960s and 1970s, a few maverick scientists considered other mate-

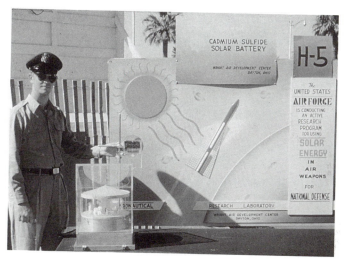

*Air Force scientists discovered and built cadmium sulfide solar cells at about the same time that silicon solar cells were developed at Bell Laboratories. This Air Force display "shows" cadmium sulfide cells driving a miniature merry-go-round. However, the Air Force failed to mention that it had surreptitiously attached the carousel to two small wires that had been carefully buried underground and connected to a well-hidden power outlet. Their cells had degraded very quickly when exposed to the elements and did not work.*

rials. Elliot Berman had initially interested Exxon in the possibility of making a low-cost solar cell from dyes which convert sunlight into electricity, an idea that has recently resurfaced in a modified form. Berman envisioned depositing the dyes onto a continuous roll-to-roll material, much like the production techniques used in the photographic industry. Exxon gave up on this approach when its researchers failed to achieve conversion efficiencies high enough to be commercially attractive.

A group at RCA, under the direction of David Carlson, also tried its hand at making a cheaper solar cell. The oil embargo of 1973 and the subsequent quadrupling of petroleum prices had triggered Carlson's interest in solar electricity. Initially he tried depositing a thin layer of polycrystalline silicon onto common materials, like glass, as Shurland had done with cadmium sulfide. When exposed to light, some of the films he made showed a small but significant photovoltaic effect. The RCA group was so sure that Carlson had successfully made a thin-film form of polycrystalline silicon that Christopher Wronski, the scientist in charge of measurements, based the calculations for his conclusions on the properties of polycrystalline silicon. This led to "getting ridiculous results . . . impossible numbers," Wronski recalled.[26] When the X-ray analysis of the supposed polycrystalline film came back to the lab, Wronski discovered why he had erred. "The X-rays showed no evidence of crystallinity," Carlson said. "Instead, it revealed that I had made an amorphous silicon solar cell [silicon which lacks the ordered internal structure inherent in crystalline silicon]. I just stumbled into it."[27]

Two years later, in 1976, after much fine-tuning, Carlson and his coworkers had upped the efficiency of their new solar device from under 0.2 percent to a significant 5.5 percent. They reported the accomplishment in the journal *Applied Physics*. The prestigious weekly *Science* called the work at RCA "[p]erhaps the most intriguing recent development" in the solar cell field.[28] The announcement stirred much excitement throughout the world, because, as David Carlson wrote, "These cells have the potential of producing low cost power since inexpensive materials such as steel and glass can be used as substrates." The relatively high efficiency surprised many scientists because the prevailing theoretical models had predicted, according to Carlson, "that you couldn't make good solar devices out of amorphous materials." Accepting the consensus "that solar cells made with [amorphous silicon] could never have efficiencies of more than 1 percent," many in the scientific community had ignored it.[29] "Fortunately," wrote David Adler, a scientist familiar with what went on at RCA, "this logic did not deter Carlson and his co-workers."[30]

The ensuing flurry of amorphous silicon research activity worldwide revealed why Carlson and his group had succeeded where earlier investigators had failed. Prior to RCA's initial work, most experimenters investigated only pure amorphous silicon, whose totally disordered composition makes the material useless as an electronic device. By inadvertently contaminating the amorphous silicon with hydrogen, Carlson and his team had greatly improved its photovoltaic capabilities. The hydrogen acted like a molecular repairman. It cleaned up the disheveled internal structure of the pure amorphous silicon by forming chemical bonds with the dangling silicon atoms. "The 'best' amorphous silicon" turned out to be "a silicon-hydrogen alloy," commonly called hydrogenated amorphous silicon.[31]

As the RCA group delved deeper into hydrogenated amorphous silicon's behavior, it appeared that the material had a self-destructive trait. This was discovered after Wronski measured a sample exposed to the sun, his co-worker David Staebler retested the sample a month or so later, and their measurements did not agree. "Then we got into an argument as to who measured correctly," Wronski remembered. "We finally discovered that both of us did the measurements right and we found that there was something very wrong with this device—it degraded in sunlight." This revelation gave the RCA people a good scare. "We were very concerned about it," Wronski admitted.[32]

*David Carlson preparing one of the first batches of "hydrogenated-amorphous silicon" at RCA Laboratories.*

Further studies eased their fears. It was learned that although hydrogenated amorphous silicon cells degrade during their first few months in the sun, they then stabilize. "It was not a catastrophic failure. It was something predictable," Carlson recalled. The cells were like jeans: You buy them a little large because they shrink a certain predetermined amount after the first washing, but they don't get any smaller with subsequent launderings. Even better news came from the lab of Joe Haneck, a member of Carlson's research team. "Joe discovered that if you make the cells very thin, the degradation became less," Carlson said.[33] A very thin cell, a hundred or so times thinner than crystalline, also significantly reduces the amount of silicon starting material needed.

RCA's resolution of the degradation problem and success in increasing the efficiency of laboratory-built cells to a record 10 percent by 1982 heightened interest in hydrogenated amorphous silicon.[34] It replaced cadmium sulfide as the expected successor to the labor- and energy-intensive crystalline silicon solar cell. Hundreds of millions of dollars were spent each year in research, development, and commercialization efforts, and novel recipes for a better solar device resulted. The addition of carbon helped; adding more hydrogen dilution in the gases used to make the amorphous silicon alloy improved things, too. The development of a multilayer cell contributed to higher efficiencies and greater stability.

In the early 1980s, Dr. Subhendru Guha, a scientist at United Solar Systems Corporation (USSC), a manufacturer of hydrogenated amorphous silicon modules, stacked three subcells to form the most efficient amorphous solar device ever built. Each layer responds to a different portion of the spectrum. The bottom layer, made of silicon and germanium, takes in red light. The middle, with less germanium, absorbs green light. The top, containing no germanium, captures blue light.[35] "The spectral-splitting approach allows more absorption of the solar spectrum," Guha reported.[36] The commercial model, the triple-junction cell, which has been on the market since 1997, converts 8 percent of incoming light into electricity—a record for amorphous material, but still less than half as efficient as its crystalline competitors.

On the other hand, unlike crystalline silicon, amorphous solar cells gain power as temperatures increase, which is ideal both for rooftop placement and the very hot climates common to the developing world. The triple-junction cells' peculiar ability to perform well under low-intensity sunlight makes them a good choice for cloudy areas, such as much of Asia.

Guha also developed a unique manufacturing process. Machines deposit his recipe onto a moving line of stainless steel in a roll-to-roll process that mimics the making of newsprint.[37] This creates a light, yet very durable module, less prone to breakage and easier to transport and install. Two plants—one in the United States, the other in Japan—manufacture amorphous modules this way.

The mass use of amorphous silicon for transistors has also helped move the technology forward. For example, amorphous transistors are used in the liquid crystal displays of portable computers; they are also widely used in laser printers and X-ray machines. "These commercial applications have sustained the drive [for research and development] on amorphous silicon," Wronski believes. "They have helped with the research and have solved a lot of the manufacturing problems crucial to success in a factory environment."[38]

A thin-film amorphous photovoltaics industry has gradually evolved from laboratory to factory. After limited production for more than a decade, a number of companies, including Solarex and USSC, have built plants capable of producing many megawatts of hydrogenated amorphous silicon solar modules. "Mass production is starting," Wronski observed, "so the next couple of years will tell" if amorphous silicon will attain its promise as the low-cost solar cell that will move photovoltaics into the mainstream power market.[39]

While government and private funding during the early 1980s went primarily to those working on amorphous silicon and ingot- and ribbon-grown polycrystalline silicon, a few researchers, such as Dr. Allen Barnett, a professor of electrical engineering and former director of the Institute of Energy Conversion at the University of Delaware, struck out on their own to develop other photovoltaic materials. In 1980, Barnett found himself at the crossroads of his professional career. He had done research on the cadmium-sulfide solar cell for the university, but because the cells' stability problems had not been solved, money from government and private industry began to dry up. This situation led him to wonder if he had a future in photovoltaics or if he ought to change fields. To help make the decision, Barnett convinced the Solar Energy Research Institute (SERI) to fund a study to examine the prevailing wisdom of which photovoltaic materials would eventually become

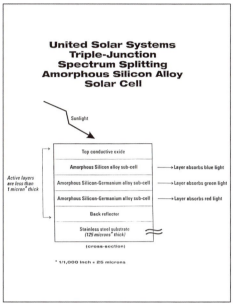

*The triple-junction spectrum-splitting amorphous silicon alloy solar cell.*

cheap enough and/or efficient enough to compete with utility-generated power. Although he agreed with the majority that the price of single-crystalline wafers would never drop low enough to challenge the bulk power market, he concluded that none of the technologies that government researchers favored stood a chance either.

For Barnett, only two possible winners existed—cadmium telluride, another photovoltaic material under consideration, and thin crystalline [polycrystalline] silicon. He then asked himself if he wanted to devote the rest of his professional life to taking one of these materials from the laboratory to the manufacturing stage. He thought first about cadmium telluride, but he could not rationalize working once again with a heavy metal for a supposedly environmentally clean power source. Dr. Richard J. Komp, a veteran in the photovoltaics field, breakfasted with Barnett "ages and ages ago" and recalls him "grousing about that." Barnett's dislike for cadmium left him to ponder thin crystalline silicon. At the time, though, "it was common knowledge that you couldn't have [a] thin crystalline silicon" solar cell, Komp said.[40] As Dr. Ting Li Chu, who had attempted, but failed, to make effective photovoltaic devices out of thin crystalline silicon, wrote almost a decade

earlier, "The . . . ineffective absorption of [light] limit[s] the use of polycrystalline silicon film for solar cells."[41] Because traditional crystalline silicon
absorbs light poorly, a relatively thick portion of photovoltaic material is
needed to retain the incoming light. This is analogous to absorbing water
with newspaper: It takes a lot of newspaper to soak up what a small sponge
can handle. Trimming the thickness of crystalline silicon invariably reduced
the amount of light it could absorb, thereby decreasing its power output.[42]
In fact, past failures to make high quality thin crystalline silicon cells steered
research toward other thin-film materials, like amorphous silicon.

Barnett felt that he needed to find a way to improve thin crystalline
silicon's light-absorbing capacity. Otherwise, the material was never going
to reach high enough efficiencies. One morning, while Barnett was looking for a parking space at the university, the period he calls "my most
creative time," he suddenly found a way out of the dilemma. "I remember
saying 'Holy cow, I have the solution!' If I roughen the surface of the
supporting material on which the thin silicon would lie, the incoming light
would bounce off at such an angle as to remain in the cell until it was
absorbed!"[43] Exxon's Research Science Laboratories quantified the effectiveness of Barnett's light-trapping idea, showing that it could force the
light to make fifty more passes inside than an equivalently sized untreated
crystalline silicon cell would allow.[44]

Radically improving crystalline silicon's ability to absorb light allowed
Barnett to consider the formerly unthinkable—trimming crystalline silicon
without drastically lowering its efficiency. Amorphous silicon researchers
found light trapping helpful, too.

Barnett soon learned that with the light absorption problem solved,
thinner crystalline silicon solar cells held many advantages over thicker
ones. First, using less material cuts manufacturing costs. Studies in the early
and mid-1980s also showed that the smaller base width of thin-film crystalline silicon decreases power losses as the temperature rises.[45] Thin-film
crystalline silicon, therefore, would be preferred in high-temperature climates and on rooftops where the heat can adversely affect the performance of thick crystalline silicon. Also, the starting material for thin-film
crystalline silicon does not have to be as high quality as that for thick
crystalline silicon because the freed electrons have significantly less distance to travel to the collecting area. They therefore have a greater chance
of reaching the contacts even under less than ideal conditions.

Just as adding hydrogen improves amorphous silicon, it can also upgrade this cheaper, lower-quality silicon. The reduced thickness also makes thin-film crystalline silicon amenable to further improvement when highly dosed with boron, another factor that gives manufacturers greater leeway in choosing a starting material.

Satisfied with the soundness of the concept, Barnett decided to stay in the photovoltaics field. He proceeded with developing a low-cost process for making inexpensive, high-performance thin-film crystalline silicon solar cells by running an inexpensive rigid material along the surface of a pool of hot liquid silicon. The silicon crystallizes to the bottom of the cool supporting material on contact.

AstroPower, the company Barnett founded to manufacture thin-film crystalline silicon, recently opened a plant capable of producing nine megawatts of the new material annually. Since only very simple equipment is required and the production of cells takes just minutes, AstroPower has already reported significant reductions in manufacturing costs compared to its thick-crystalline competition.[46]

Barnett's work did not go unnoticed. Dr. Martin Green, one of the world's most respected photovoltaics researchers, has come up with another promising thin-film crystalline silicon device, while acknowledging Allen Barnett's priority in commercially exploiting the potential of thin-film crystalline silicon.[47]

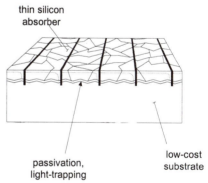

*AstroPower's thin-film silicon solar cell.*

Green is Australian and his work in solar cells goes back to the days when Telecom Australia began equipping its microwave repeaters with solar cells. "I'd be bumping into Arnold Holderness and Michael Mack at conferences since we were the only ones in Australia at the time who knew anything about photovoltaics," Green recalled. He also credited Telecom Australia's pioneering commercial commitment to solar cells for helping him become one of the premier research scientists in the field. "It obviously stimulated my work because I had a target audience . . . there was a real interest by the Telecom people in actually using these things."[48]

In the early 1980s, Green and his colleagues at the University of New South Wales focused their attention on improving the efficiency of regular crystalline silicon cells. The approach seemed to them the only way to make crystalline silicon more affordable, having concluded, "[N]o solution appears to be in sight for the problem of slicing [the starting material] cheaply."[49] The Australians felt confident they could raise the efficiency substantially, up to 20 percent, without increasing manufacturing costs, thereby providing more watts per dollar.

The performance of commercial crystalline silicon modules available in the 1970s and 1980s, Green observed, "hinged on how metal contacts were formed to the cell. Therefore, that became the area we paid particular attention to."[50] Though screen printing of contacts has streamlined production and, for this reason, has been adopted by the entire industry, the New South Wales group discovered that it worsens cell performance. Because the additives in the silver paste that make screen printing possible are less conductive than pure silver, they significantly hinder the contacts' capability to "pull" liberated electrons out of the cell. And because the relatively wide contact lines shade segments of the cell, increasing the number of contacts would obscure too much photovoltaic material from the sun. Therefore, the freed electrons have to somehow be "herded" to available collection points for the production of electricity. Adding phosphorous to the surface of the cell accomplishes this. Unfortunately, greatly increasing the phosphorus content creates a dead layer on the cell's surface; light absorbed in this inactive region is wasted since it cannot generate electricity.[51]

As an alternative to screen printing, Green and his colleague Stuart Wenham "hit upon the idea to use a laser to form grooves on the surface of the cell" and to fill the grooves with copper.[52] The copper contacts have three times the conductivity of silver paste; partially buried inside the grooves, they obscure less of the cell's surface, which allows more sunlight to reach the cell. More contacts can go on the surface without too much shading of the cell, which adds to the number of collection points inside the silicon and eliminates the need for all that phosphorus. Modules made this way have consistently outperformed all others in the world.

The Australians' ultra-efficient cells got their first test in the 1990 "World Solar Challenge," where photovoltaic-powered vehicles raced across the Australian continent. The Japanese contender, the clear favorite, was a Honda powered by modules rated as the best in the world. However, a Swiss car, which ran on modules from Green's laboratory, unexpectedly won. The victory attracted national and world attention to the photovoltaics work being done by Green and his colleagues. Two years later, British Petroleum bought the rights to manufacture the high-efficiency modules developed at the University of New South Wales, which further bolstered the prestige of Green and his colleagues. In recognition of these achievements, the Australian government awarded them a research center.

The increased funding has allowed Green's group to expand its focus. Prior money constraints had forced the Australians to concentrate their efforts on improving conventional crystalline silicon technology, though they yearned to work on thin-film cells. "We always felt that thin films held the answer to the eventual success of the technology because a highly efficient thin-film photovoltaic device could be cheap enough to provide a large fraction

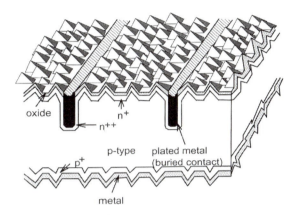

*The buried contact cell developed by Martin Green and Stuart Wenham of the University of New South Wales, Australia. The little pyramids on top help to trap light.*

of the world's power," Green asserted. "But the practical dictates of obtaining funding did not allow us to do much work on them." With money no longer an issue, work on thin-film devices began in earnest. The timing of the award fit Green's professional career perfectly. "I have fifteen good years of research left in my life," Green mused right after hearing about the honor bestowed upon his group, and "[I] need fifteen years to develop a highly efficient thin-film technology that will make significant inroads on the world's electrical markets."[53]

Green's thin-film crystalline silicon device is completely unlike Barnett's. By depositing very thin alternating layers of positive and negative silicon onto glass, all potential electrical charges are near a p–n junction, the core of any solar cell, where the important photovoltaic activity occurs. Grooves laser-cut into each stratum and filled with metal allow access to each layer and provide the freed electrons an easy route to the surface, making for an extremely efficient collection system.

Worldwide interest in the proposed cell gave Green the confidence to take the concept to Australia's utilities, which were just becoming interested in photovoltaics. He wanted to see if they would form a consortium to finance further work on the technology to the point where its commercial impact could be assessed. Green and colleague David Hogg first approached Pacific Power, the largest utility in Sydney, where the university is based. The utility found the idea so intriguing that it wanted exclusive rights to the

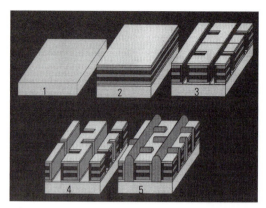

*The steps in processing Pacific Solar's thin-film crystalline silicon solar cell. (1) Its manufacture starts with a piece of glass for the base. (2) Four thin alternating layers of positive and negative silicon are deposited onto the glass. (3 & 4) A laser cuts grooves into the layers of positive and negative silicon. (5) Metallic contacts are placed in the grooves to collect the electrons propelled by incoming photons from the sun and so make electricity.*

technology. "They wanted all the action. . . . So we formed a company, Pacific Solar, which was a joint venture between Pacific Power and the university's commercial arm, Unisearch Limited," according to Green.[54]

Pacific Solar put up $50 million to take the thin-film device from the drawing board to large-scale production, making this the largest investment in renewable technology development in Australian history.[55] The utility's willingness to invest so much money without the expectation of immediate returns lured some of the best people in the crystalline silicon cell field to leave their jobs and countries to work for Pacific Solar. Dr. Paul Basore, for example, gave up a senior position at Sandia National Laboratories for what he considers the chance of a lifetime. "The problem has always been that the little government handouts have never been enough to overcome the huge technological barriers that are involved [in] developing a new solar-cell technology. Down here they have pledged millions of dollars over a five-year period, understanding how much money it takes. . . . When you see an opportunity where you have five years guaranteed in which [to] work on long-term goals before having to meet the bottom line, that signaled to me there was a real possibility to accomplish something very significant."[56]

A great advantage in manufacturing any thin-film solar product is that it requires much less energy than the production of traditional solar cells.[57] This large energy savings makes thin films a superior ecological product, significantly decreasing the amount of polluting fuels consumed in solar cell production. But using very small amounts of photovoltaic material does have its drawbacks, as Martin Green himself admits: It takes a lot more corrosion to stop a conventional crystalline wafer from working than when relying on an itty-bitty piece of photoactive matter. The question then naturally arises: Will the necessity of improved packaging bring the price up to a level that will wipe out the savings accrued in manufacturing the thin cells?

Perhaps engineers at Evergreen Solar have solved the dilemma, having found better and less costly materials to encase the cells. They have, for example, introduced a back cover "that is four times thicker, less expensive, and lasts longer" than those currently in use, according to the company's president, Mark Farber. "Best of all, we can wrap it around the panel's edges so that it actually ends up framing the module . . . eliminating the conventional aluminum frame, which, after the photovoltaic substance, is the next most expensive material."[58]

The number of potentially inexpensive ways to make solar cells being pursued is dazzling. They include sheet crystalline silicon, amorphous silicon, thin crystalline silicon, cadmium telluride, and copper indium diselenide.[59] Even traditional crystalline silicon might surprise the pundits. It still has the potential to drop significantly in price. Only time will tell which material will dominate. Or perhaps they will give each other a good run for the money. One point, though, seems beyond dispute, as a strategic planner for a major Australian utility explained: "In the 1980s, you could argue it was an issue of whether photovoltaics would get down to be competitive with grid electricity. By the early 1990s, you had to say it was but a question of when."[60]

## Notes & Comments

1. J. Caldwell, "Photovoltaic Technology and Markets," *Contemporary Economic Policy* 12 (April 1994): 101–2. The cost per kilowatt hour of electricity produced by a photovoltaic module depends upon (1) the cost per peak watt installed; (2) the longevity of the module and ancillary equipment; (3) the amount of electricity produced by the module; and (4) the terms of financing the installation.
2. *Photovoltaics for Municipal Planners*, NREL/TP-411-5450 (April 1993), 14, 15.
3. D. Osborn, "Photovoltaics," in *Solar Energy: Today's Technologies for a Sustainable Future* (Boulder, CO: American Solar Energy Society, 1997), 17, 19.
4. Interview with Steve Taylor of Edison Technology Solutions.
5. Interview with John Thornton.
6. Interview with Bob Johnson.
7. D. Chapin, "The Direct Conversion of Solar Energy to Electrical Energy" (1956?), 30. (Courtesy Audrey Chapin Svensson.)
8. D. Chapin, "Progress Report for May and June 1955," 1 July 1955, Dept. 1314; D. Chapin, "Progress Report for May and June 1954," 1 July 1954, Dept. 1314. (Courtesy Audrey Chapin Svensson.)

9. Interview with Fritz Wald.

10. Interview with a photovoltaics pioneer who wishes to remain anonymous.

11. M. Wolf, "Photovoltaic Solar Energy Conversion Systems," in J. Kreider and F. Kreith, *Solar Energy Handbook* (New York: McGraw-Hill 1981), 24–25.

12. M. Savelli, "Thin-Film Solar Cells and Spray Technology," *Solar Cells* 12 (1984): 192.

13. Interview with Fritz Wald.

14. P. Rappaport, "Working Group and Discussions," *Photovoltaic Conversion of Solar Energy for Terrestrial Applications—Workshop Proceedings*, N74-22704 (23–25 October 1973) (Cherry Hill, NJ: Jet Propulsion Laboratory, 1974), 10.

15. G. Schwitthe, "Some Comments on Ribbon Growth of Silicon," *Photovoltaic Conversion of Solar Energy for Terrestrial Applications—Workshop Proceedings*, N74-22704 (23–25 October 1973) (Cherry Hill, NJ: Jet Propulsion Laboratory, 1974), 32.

16. A. Mlavsky, "U.S. Photovoltaic Systems," *Japanese/United States Symposium on Solar Energy Systems*, Summaries of Technical Presentations (Washington, D. C., 3–5 June 1974) (Washington, D.C.: MITRE Corporation, 1974?), 8-4.

17. A. Taylor et al., "Long Nonagons—An Approach to High Productivity Silicon Sheet Using the EFG Method," *Journal of Crystal Growth* 82 (1987): 134.

18. Interview with Fritz Wald.

19. Taylor et al., "Long Nonagons."

20. Mlavsky, "U.S. Photovoltaic Systems."

21. Interview with Mark O'Neill. The demand for concentrator systems remains very small because most current purchasers of solar cells are either remote industrial customers or residents of the developing world who cannot risk breakdowns or tolerate the level of maintenance that concentrators require.

22. D. Chapin, "The Conversion of Solar Energy to Electrical Energy," UCLA Lecture Notes, 18 May 1956, 18. (Courtesy Audrey Chapin Svensson.)

23. F. Shurland, "The History, Design, Fabrication and Performance of CdS Thin-Film Solar Cells," in C. Backus, *Solar Cells* (New York: IEEE, 1966), 66.

24. "Sun-Power Prospects," *Chemical Week* (1 April 1967): 53.

25. NASA Lewis, *Photovoltaic Conversion of Solar Energy for Terrestrial Applications—Workshop Proceedings*, N74-22704 (23–25 October 1973) (Cherry Hill, NJ: Jet Propulsion Laboratory, 1974), 23.

26. Interview with Christopher Wronski.

27. Interview with David Carlson. Disagreement exists over who discovered the amorphous silicon solar cell. Some have steadfastly maintained that the credit goes to Walter Spear and Peter Le Comber, who did their work at the University of Dundee in the United Kingdom. The United States Patent Office, however, recognizes David Carlson as the inventor of solar cells made from amorphous silicon (U.S. Patent No. 4,064,521). Several court battles have upheld Carlson's priority.

28. H. Kelly, "Photovoltaic Power Systems: A Tour through the Alternatives," *Science* 199 (10 February 1978): 637.

29. D. Carlson et al., "Solar Cells Using Schottky Barriers on Amorphous Silicon," *The Conference Record of the Twelfth IEEE Photovoltaic Specialists Conference* (Baton Rouge, LA) (New York: IEEE, 1976), 893.

30. D. Adler, "Amorphous Silicon Solar Cells," *Sunworld* 4, no. 1 (1980): 17.

31. A. Robinson, "Amorphous Silicon: A New Direction for Semi-Conductors," *Science* 197 (25 August 1977): 852.

32. Interview with Christopher Wronski.

33. Interview with David Carlson.

34. D. Carlson, "The U.S. DOE/SERI Amorphous Silicon Research Project," *The Conference Record of the Eighteenth IEEE Photovoltaic Specialists Conference* (Las Vegas, 21–25 October 1985) (New York: IEEE, 1985), 474.

35. United Solar Systems Corporation, "United Solar Systems Corp. Technical Profile," February 1996.

36. Interview with Dr. Subhendru Guha.

37. S. Guha, "Cost-Effective Electricity from Amorphous Silicon Alloy," *Photonics Spectra* 30 (July 1995): 112.

38. Interview with Christopher Wronski.

39. Ibid.

40. Interview with Dr. Richard Komp.

41. T. Chu, "Polycrystalline Silicon," *Photovoltaic Conversion of Solar Energy for Terrestrial Applications—Workshop Proceedings*, N74-22704 (23–25 October 1973) (Cherry Hill, NJ: Jet Propulsion Laboratory, 1974), 17.

42. C. Sah et al., "Effect of Thickness on Silicon Solar Cell Efficiency, *IEEE Transactions on Electron Devices*, vol. Ed-29, #5 (May 1982): 903.

43. Interview with Allen Barnett.

44. E. Yablonovitch and G. Cody, "Intensity Enhancement for Textured Optical Sheets for Solar Cells," *IEEE Transactions on Electron Devices*, vol. ED-29, #2 (1982): 300.

   U.S. Patent No. 4,571,448 recognizes Allen Barnett as the developer of the back reflector on a substrate to enhance solar cell performance.

45. J. Knoblock et al., "High Efficiency Crystalline Thin Film Solar Cells with Diffuse Reflector for Optical Confinement," *Fifth E.C. Photovoltaic Solar Energy Conference* (Athens, Greece) (Dordrecht: Kluwer Academic Publishers, 1984), 285.

46. S. Vaughn, "Firm Finds Better Way to Make Solar Cells," *Investor's Business Daily* (7 April 1999): reprint.

47. M. Wald, "New Design Could Make Solar Cells Competitive," *New York Times* (4 June 1994): C9.

48. Interview with Martin Green.

49. M. Green, "Solar Cells: Future Directions," *Solar Cells* 12 (1984): 95.
50. Interview with Martin Green.
51. Interview with Stuart Wenham.
52. Ibid.
53. Martin Green, 1994, videotape interview. (Courtesy Mark Fitzgerald.)
54. Interview with Martin Green.
55. "Pacific Solar Annual Report" (1997), 3.
56. Interview with Paul Basore.
57. K. Srinwas, "Comparative Study of Photovoltaic Technologies for Large Scale Terrestrial Applications," *Electronics Information and Planning* (Bombay: Information Planning and Analysis Group of the Electronics Commission, 1995), 399.
58. Interview with Mark Farber.
59. The scarce supplies of cadmium, germanium, indium, and tellurium may impose strict limits on the production of solar cells made from them. It is also possible that the residues from the large-scale use of these materials could prove toxic. However, amorphous silicon (when not combined with germanium) and single and polycrystalline silicon will never face supply concerns, no matter how large the demand. Nor will the manufacture or use of these cells (with or without germanium) pose any threat to the environment. B. Andersson et al., "Material Constraints for Thin-Film Solar Cells," *Energy* 23, no. 5 (1998): 407–10.
60. Interview with Peter Lawley, an engineer for Pacific Power and an officer in Pacific Solar.

Chapter Fifteen
# The Silent Revolution Continues

Twenty-six years of terrestrial experience have truly created a place for photovoltaics in the world energy field. "Two decades ago PV for uses on earth was completely new for everything and everybody," Bernard McNelis observed. "Back then most people had no idea that there was such a thing. Now, at least it is known by people who do science and engineering."[1] Indeed, industry no longer considers photovoltaics an alternative source of energy for its remote power needs. It is now regarded as "the most effective solution."[2]

Back in the 1970s, western institutions, such as the World Bank, did not even consider photovoltaics for their current or future energy programs. As former World Bank employee Steve Allison affirms, "The Bank's attitude toward photovoltaics [then] was nonexistent."[3] Throughout the 1980s, nothing changed. "It was virtually impossible, [even] at your own cost, to go to the World Bank and put on a one- or two-hour lecture merely to inform them about photovoltaics," Terry Hart, now a consultant for World Bank solar projects in India, confirmed.[4] The inertia at the Bank and like institutions toward embracing photovoltaics was "less a reflection on a lack of scientific or technological" merits of solar cells, the internationally respected scientific journal *Nature* believed, "than on the enormous built-in momentum that modern industrialized society has which resists major changes."[5] Bureaucracies shy away from anything but the tried and conventional. In Allison's opinion, every World Bank functionary worries that "he or she might suggest something that will turn out wrong, and God knows what might happen to your career after that!"[6]

The first members of the energy establishment who publicized successful earth applications were the oil companies and the Electric Power Research Institute (EPRI), the research arm of America's investor-owned utilities. Early in 1978, Shell issued a report to the public on solar energy, announcing that the sun "may provide electric power via photovoltaic devices in certain remote applications . . . where inaccessibility otherwise makes equipment operation and maintenance involving other energy sources extremely costly."[7] EPRI clearly saw the worth of terrestrial photovoltaics during the 1970s and early 1980s, too. "Power has a very high value in . . . remote applications, and solar cells generate it more cheaply than any other means," the Institute wrote in 1981.[8]

Remote industrial applications—such as powering navigation aids, corrosion protection devices, railroad signaling apparatuses, water pumping, and home power systems—might have seemed trivial to those who think in terms of central power generation, but, as one expert observed, "Benefits do accrue . . . in terms of field testing, user education, and confidence."[9] The success of photovoltaics in powering remote microwave repeaters erased any doubts engineers might have had about its reliability. Telecommunications professionals would attend international conferences and spread the word that photovoltaic systems perform better than any other stand-alone power generator. Eventually such praise reached institutions such as the World Bank and the World Energy Council, an organization that represents major utilities throughout the globe, leading people like Dr. Hisham Khatib, a member of the Council's Committee for Developing Countries, to recognize that "solar cells for use at individual houses . . . are a very important development that warrants particular attention [as] they are ideal for low-power rural applications."[10]

Both Steve Allison and Terry Hart have noted the World Bank's recognition of the importance of photovoltaics for the developing world. On a return visit to the Bank, after a twenty-five-year absence, Allison could not believe the number of people involved in photovoltaics. "There's a whole department with specialists working away on the deployment of solar cells," Allison noted.[11] Hart observed, "The Bank portfolio has been

*Thousands of Mongolians provide electricity to their yurts via solar modules. Photovoltaic technology has allowed these nomadic people to continue their wanderings while enjoying the amenities that electricity can provide.*

rapidly growing in terms of interest and commitment" to photovoltaics.[12] A recent statement made by the Bank declared, "Few people now doubt that PVs have an important and growing part to play in providing electrical services in rural areas of the developing world, and many are also becoming aware that PVs have potential applications in suburban and peri-urban areas of many developing countries."[13]

The continuing revolution in telecommunications will also bring an increased role for photovoltaics. The development of solar cells has always gone hand-in-hand with advances in telecommunications. Laying the trans-Atlantic telegraph cable brought about the discovery of selenium's light sensitivity and, ultimately, its ability to convert sunlight directly into electricity. Transistor research gave birth to the silicon solar cell, still the workhorse of photovoltaic applications in space and on earth. Subsequent advances in the telecommunications field would literally have never got off the ground without solar power. Ever since

*A soldier with a photovoltaic military trifold panel.*

the launching of Telstar in 1962, the transmitters of all telecommunications satellites have run on photovoltaics. Solar cells give satellites the power to beam telephone calls, Internet messages, and TV shows to relay stations back on earth for worldwide dissemination.

The union of photovoltaics with the communications industry has continued terrestrially as well when, in the 1980s, solar cells became the power source of choice for microwave repeaters. The spread of cellular networks continues this close relationship. Jim Trotter relates how his firm, Solar Electric Specialties, got involved in the field: "When we were first doing photovoltaics in the early 1980s, cellular phones were a rare and exotic breed, the cell sites were fewer and more concentrated. They were usually located on mountaintops that had been used for decades by other broadcasting groups where there was utility power available. But as the cell sites sprawled into virgin territory and there was no power available, the cost to extend power lines to these sites tends to exceed the cost of solar cells in lots of locations, making us cost-competitive in many cases for a technology that wasn't there when we started out."[14]

Most westerners take telephone access for granted. In Africa, however, 75 percent of the telephones are in the cities, while more than 75

percent of the population reside in rural areas.[15] Many Africans, Asians, and Latin Americans—in fact, over half of the world's population—must travel over two hours just to make a phone call.

As with electrical service, the expense of stringing telephone wires keeps most of the developing world isolated. Satellites, cellular service, or a combination of the two offer the only hope. In the Dominican countryside, for instance, "cellular is the only phone service you can get," according to Richard Hansen. "We're now seeing phone companies putting in pay phone booths with solar cells . . . fixed cellular, not like the portable ones in the States." Solar-powered cellular phones in rural grocery stores or restaurants "also make a good business by creating a little cellular phone

*Solar-powered satellites, like Motorola's Iridium®, make it possible to operate photovoltaic pay phones in places where wires will never go.*

calling center," Hansen adds.[16] The Grameen Bank in Bangladesh has similar plans for developing a photovoltaic-powered cellular network. Through its subsidiary, Grameen Telecom, it will finance fifty thousand solar-powered cellular phones for one million subscribers. The owners of the phones, impoverished villagers, will first pay off the equipment and then earn money by charging others to make calls.[17]

A group of influential investors, including Bill Gates and major telecommunications companies, have formed a commercial venture called Teledesic, which plans to bring the Internet to every village in the world. Aided by photovoltaic-run low-orbiting satellites communicating with photovoltaic-powered land-based satellite receiver dishes, artisans, for example, could communicate on their photovoltaic-powered laptops directly to customers anywhere in the world. Customers could examine and select items displayed on websites, and they could place orders and make payments electronically. Such improved communications would rid the artisans of their dependency on urban middlemen.[18]

The global Internet would also allow villagers, if properly trained, to work at relatively high-paying jobs without leaving home. With such training and equipment, "Young people will be able to perform data entry and . . . transcription services for any company in the world [without leaving their villages], a better alternative than migration to urban slums in search of employment," states Muhammed Yunus, founder and managing director of the

*Solar-powered Iridium® communication satellites link solar-powered phone booths in the developing world.*

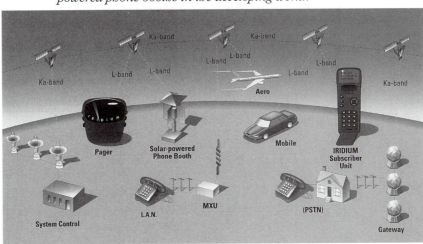

Grameen Bank.[19] As a United Nations' study concluded, "The use of solar photovoltaics . . . can revolutionize [communications in the developing world] by, for the first time, offering a real practical possibility of reliable rural telecommunications for general use."[20]

Opportunities for photovoltaics in the developed world continue to grow, too. The U.S. National Park Service, for example, has finally decided that the electricity produced by its many generators not only costs more than power generated by photovoltaics, but the noise of the generators, the pollutants they spew, and the risk of oil spills in transporting the diesel fuel over waterways run counter to the agency's mission as the guardian of America's pristine landscapes.[21] The Defense Department has also identified over three thousand megawatts of power now produced by diesel generators that photovoltaics could more economically generate. This alone is thirty times the capacity of today's solar cell industry.[22]

*An increasingly common sight: mobile highway signs powered by photovoltaics.*

Private industry has also begun to replace its generators with photovoltaics. For years, the portable warning signs used on roadways to alert motorists of lane closures and other temporary hazards have run on gasoline generators. "But the maintenance was so bad," a battery engineer attests, "that they started to use solar."[23] In fact, the entire industry is changing to photovoltaics as fast as the necessary equipment can be brought on-line. Revolutionary lighting elements, light-emitting diodes, have helped spur the replacement of generators with solar power. The diodes require so little power that a small panel can not only power the lights during sunny periods but also charge accompanying batteries to keep the lights running at night and during bad weather.

In the United States and western Europe, hundreds of thousands of remotely situated homes are not connected to power lines. Just as in the developing world, linking these homes to centrally generated electricity costs too much. For example, Public Service of Colorado, one of the state's largest utilities, requires customers to pay tens of thousands of dollars to string wires one-quarter to one-half mile (four to eight hundred meters) from the closest line. "Once these people pay to be wired, they can get the electricity quite cheaply," John Thornton observes. "But first they have to

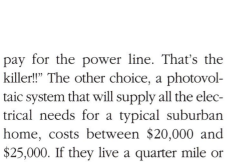

*Vacationers, like those who inhabit these Swiss chalets, will never have to worry about power lines obliterating their rural setting. All of their electricity comes from the solar modules mounted on the rooftops.*

pay for the power line. That's the killer!!" The other choice, a photovoltaic system that will supply all the electrical needs for a typical suburban home, costs between $20,000 and $25,000. If they live a quarter mile or more from a utility pole, Thornton concludes, "people are finding that solar cells are the lowest cost option."[24]

The electrical requirements of the thousands of vacation cabins in the developed world closely resemble those of rural homes in poorer countries. Vacationers tote in either kerosene, diesel, or gasoline to run a few lights and perhaps a TV or radio. "These are the people who can most beneficially use photovoltaics," Jim Trotter believes. "That's because a small solar system costing a few hundred dollars, which they can easily afford, can satisfy all their needs."[25] As early as 1981, ARCO Solar found a good market in Spain, supplying photovoltaic power to the mountain retreats of those escaping the hot, muggy Mediterranean summer. Because the private utility could not electrify them without losing money, these mountain homes had no power. During the 1980s, Spaniards bought more photovoltaic systems than anyone else in Europe. Some years later, powering vacation homes in Finland put that nation's utility into the photovoltaics business. The utility decided it made more sense to install solar modules on each of the forty thousand holiday cottages in the Finnish countryside than to bring in power lines that would desecrate the bucolic vistas. For similar reasons, fourteen thousand Swiss Alpine chalets get their electricity from the sun.

Roland Skinner, general manager of the Northwest Rural Public Power District, a Nebraska rural electric cooperative, knows that Mali and the rest of developing world are not the only places that use water pumps. His utility serves the picturesque but thinly populated Sand Hills region. Much of the electricity that runs through Skinner's lines powers water pumps, and because the cooperative's clients have low energy needs, the organization has found itself in a dilemma. In the next decade, the rural utility will have to replace its aging sixty-year-old poles. Skinner figures that for "the longest connections, the old ones," which run about a mile and a half [approximately 2.5 kilometers] to drive a water pump, it will cost "about $15,000 to put in new poles and wire and $5,000 to $6,000 to tear the old ones out. . . . Or I could put in a photovoltaic system for $5,000 total!" The economics for powering wells with solar cells become even more compelling, according to Skinner, "with the continuing disappearance of family-owned farms. A lot of homesteads are falling down and the only demand for electricity left out there is a well for the livestock."[26]

Managers of other rural utilities have similar concerns. They, too, own miles of dilapidated line which principally power stock wells—about 125,000 of them. Another 125,000 water pumps in the western United States run on diesel or wind, and all of them, in Skinner's words, are "getting older by the year."[27] Environmental concerns will only increase the number of pumps in the future. Missouri, for example, wants to move cattle away from its streams and rivers, according to Ken Stokes, who runs a photovoltaic equipment buying service that caters to rural Americans. "Livestock are in there and doing their thing [defecating] in the water and they're also messing up the embankments and eroding them. The state therefore wants to fence off all waterways and have cattlemen water their animals by pumps."[28] The number of water pumps will also increase as ranchers embrace the concept of rotational grazing, which calls for fencing a pasture into portions and grazing animals more intensely for shorter periods of time in each portion. Under these conditions, grass grows faster and healthier, but ranch-

*Livestock in the Sand Hills region of Nebraska mill around a reservoir of water filled by a solar-run pump.*

ers have to add pumps because the cattle must have water in each field. They could choose a diesel- or propane-fired generator to run their pumps, but then, "You've got to fuel them up, turn them off, and check the oil at least once a day. [And] if you're not there to shut them off once they pump the well dry, they can burn themselves out," according to rancher Donald McIvor. This left him the choice of either photovoltaics or wind power. Photovoltaics won hands down in McIvor's opinion, "It's trouble free and a lot cheaper."[29]

Increased ecological awareness is a big reason behind the popularity of equipping recreational vehicles (RVs) with solar modules. A good many camp-grounds have come to prohibit RV owners from running diesel generators because the exhaust gases pollute and the noise keeps other campers from sleeping or communing with nature. RVs already have inverters to change the generator's DC electricity to AC and batteries for storage, so all the industry has to do is substitute a new power source—photovoltaics.

*More and more RVers are replacing noisy diesel generators with silent solar power.*

The movement for cleaner air has led many urban and suburban dwellers to want "green" power for their homes. Market research in Australia has shown that tens of thousands of people are willing to pay a premium to power their homes with nonpolluting sources of energy. The trend toward deregulation in the electric industry will make it easier for more people to make the choice. Having a photovoltaic system on your roof "is about as green as you can get," remarks Peter Lawley, who united Martin Green's University of South Wales group with Pacific Power, thus creating Pacific Solar. "The beauty of it is when people have the modules on their rooftops and can see their meter running backward," Lawley adds. "That's real."[30] In the opinion of an American utility, which is investing in rooftop systems designed by AstroPower, "There's going to be a lot of different ways people will get electricity in the future and only one will be through traditional wires and poles. As photovoltaics becomes more cost-effective and more reliable, more and more people will be choosing [it] as an alternative to the traditional supply line."[31]

*A prepackaged stand-alone photovoltaic unit lights this sign at night.*

To make rooftop systems as simple as possible, solar technologists have come up with the AC-module, which contains a tiny mechanism that converts the direct current produced by the cells into the more commonly used alternating current before the electricity exits the panel. The AC-module eliminates both the special wiring necessary to couple the panels as well as the costly inverter previously required to change DC into AC. The homeowner simply plugs the electrical cord from the AC-module into a conventional electrical socket. *Popular Science* calls the AC-module "a significant step forward for photovoltaic technology" and included it in the magazine's 1998 list of "The Year's Greatest Achievements in Science and Technology."[32] DC systems have also become easy to install. They are now available in prepackaged stand-alone power units which purchasers simply plug in where electricity is needed.

Global warming ranks as the principal concern motivating the push toward greener energy sources. Indeed, in 1989, the *Annual Review of Energy* called the "possibility of global climate change arising from fossil fuel combustion . . . [the] one environmental issue [that] shadows all of our thoughts about energy today."[33] "Most people in the know see regulations on the output of $CO_2$ as a 'when' rather than an 'if,'" reports Mark Trexler, whose consulting firm specializes in global warming issues. In fact, the 1998 Kyoto Protocol on Climate Change, which has yet to be ratified, commits the developed world to institute significant reductions on greenhouse gas emissions. "Caps on $CO_2$ emissions could have a huge effect on renewables," according to Trexler. "Once companies need to start offsetting their emissions, there could be billions of dollars flowing" yearly from these companies to $CO_2$ offset projects.[34] And photovoltaic installations, as offset projects, would definitely receive some of the funding. For example, abiding by carbon dioxide restrictions, a utility could continue to burn fossil fuels if it financed a solar cell installation that would eliminate the emission of a corresponding amount of $CO_2$ elsewhere.[35]

To stop greenhouse gas emissions where they start—at the smokestacks, plant owners would have to remove carbon dioxide the same way utilities currently scrub sulfur dioxide. These added costs would eliminate the price disadvantage photovoltaic-generated electricity faces when competing in the utility market. The World Energy Council predicts that by adopting such a program, "[T]he impact on the attractiveness and penetration of renewable energy" technologies such as photovoltaics would be "very large."[36]

*Modules integrated into the roof of a parking structure provide both power and protection for electric vehicles.*

Photovoltaics could also play a significant role in helping to eliminate greenhouse emissions by cars and trucks, if and when electric vehicles start to proliferate. Modules integrated into the roofs of covered parking lots would provide conveniently located recharging stations, as well as protection from the elements.[37]

"This winter's El Niño [1997–98] is a taste of what we might expect if the earth warms as we now project," James Baker of the National Oceanic and Atmospheric Administration announced.[38] His warning suggests that the future will bring more calamitous weather: deluges, droughts, hurricanes, tornadoes, and ice storms, resulting in wildfires, floods, and other related misfortunes. The expected increase in the number of natural disasters brought on by the harsher future climate, as well as the growing number of people settling in catastrophe-prone regions where earthquakes, hurricanes, and volcanoes are a threat, make early-warning systems essential. Photovoltaic-powered monitoring devices in relatively remote locations can pick up signals of impending disaster and alert the population. Solar-run equipment already keeps tabs on the water flow in the canyons above Tucson, ready to warn people downstream of the flash floods that sometimes occur during their monsoon season. Railroads, such as the Burlington Northern Santa Fe, use photovoltaics to power rock and mudslide detection fences, which electronically inform trains of potentially dangerous conditions and so prevent life-threatening and costly derailments.[39]

The ultimate early-warning device may be solar-powered weather surveillance airplanes, scheduled for takeoff sometime in the next decade. The hundreds of photovoltaic panels that cover the plane's 250-foot wing-

*Prototype of a slow-flying, ultrahigh-altitude, long-duration solar-powered aircraft that could monitor the development of potentially destructive weather events. It was designed and built by AeroVironment and funded by NASA.*

spread are connected to fuel cells underneath. Throughout the day, photovoltaics will generate the electricity to run the aircraft and to extract hydrogen and oxygen from the water discharged by the fuel cells the night before. When the sun sets, the extracted hydrogen and oxygen will power the fuel cells, generating the energy that keeps the aircraft aloft at night. Water discharged in this process will allow the diurnal cycle to begin anew the next morning. In this way, the airplane can remain above the turbulence for months, watching for and tracking hurricanes and other potentially dangerous weather disturbances.[40]

Once a natural disaster strikes, power lines topple. But those who own photovoltaics with storage will still have electricity. For example, after the great ice storm that hit the American northeast in early 1998, the lives of people who powered their homes by solar cells connected to batteries were not drastically altered. Hurricanes Georges and Mitch wreaked terrible havoc throughout the Carribean and Central America, but they left the thousands of photovoltaic installations there virtually unscathed. Of the more than nine thousand installations in the Dominican Republic, fewer than twenty were lost. Similar low losses of photovoltaic systems were reported in Honduras as well. Amazingly, even in areas that took a direct hit, such as the coastal village of Bayahibe, few solar modules were lost because their owners "simply removed [them] from the roof" before the hurricane arrived and remounted them when the storm left.[41]

Catastrophes that break down the vast and complex infrastructure that makes modern life possible leave those without an alternative such as photovoltaics to fall back on as isolated and without services as the poorest, most remote villagers in developing lands. Only stand-alone equipment run on locally available power sources can bring a semblance of normality to such situations. Disasters create the ideal environment for photovoltaics: No power from the outside is forthcoming and only small amounts of energy are necessary to run basic emergency equipment. Emergency personnel can quickly carry in—by foot, if necessary—ultralight panels, which fold and fit into a backpack, and set them up in minutes to reestablish the communication links imperative to search and rescue work. Photovoltaics can also power portable message signs and temporary warning lights to help motorists negotiate dangerous roads, telling them of closures and conditions ahead and taking the place of downed traffic signals and fallen street signs.

As the aftermaths of Hurricanes Andrew and Mitch demonstrated, utility power can remain down for some time. Maintaining simple health and sanitation standards then becomes a primary concern. In such situations, photovoltaics can stem the onslaught of disease. For example, in Honduras, photovoltaic-powered ultraviolet rays penetrate contaminated

*These fabric-like folding modules, built by United Solar Systems Corporation, make ideal power packs for emergency use. In such situations, several watts of power could make the difference between life or death. The Northwest American Expedition team, at right, used the modules as their sole source of power for on-the-mountain communications during its successful ascent of Mount Everest.*

*Shell Renewables plans to market solar electric home kits through the many service stations its parent company, Shell Oil, owns in the developing world. The kits include a module, lamps, battery, wiring, charge controller, distilled water, and an instruction manual. Offering a complete system in one convenient package makes the sun's energy more accessible to more people.*

water, killing all biotic pathogens. Modules have also proven their indispensability by powering clinics and aid centers. At St. Anne's Mission, which served the victims of Hurricane Andrew, it took but one day to put up a photovoltaic system that ran fans, lights, and vaccine refrigerators.[42] The mission's rector remarked to those seeking aid and shelter, "You see? This is God's light."[43]

Just as El Niño has forewarned us of global warming's possible impact, the oil crisis of the 1970s provided the Western world with an idea of what life will be like when cheap oil runs out. Petroleum analysts Colin Campbell and Jean Laherrére predict that petroleum prices will rise once again "within the next decade," but this time they will continue to climb. Campbell and Laherrére have come to this conclusion because "the supply of conventional oil will be unable to keep up with demand" as worldwide production starts to decline by 2010.[44]

Shell Oil, like Campbell and Laherrére, has done some scenario planning based on the supply of and demand for oil, and it has come to the same conclusion, differing only in timing. Shell believes that "fossil energy sources may peak around 2020, while the global energy demand may triple by then." The European-based oil company concluded, "Renewable energy sources need to cover a substantial part of the deficit that fossil fuels cannot supply. And within renewables, photovoltaics must play a major role."[45]

To prepare for the new era, Shell Oil instituted its Shell Renewables division in 1997, which soon expects to produce forty-five megawatts of solar cells annually, equal to about 40 percent of the world's 1997 production. "When a major oil company effectively projects the end of the fossil fuel age," *Scientific American* told its readers, and aggressively enters the photovoltaics market, "it is a sure harbinger . . . [of] the future."[46] With the anticipated mandatory global curbs on fossil fuel emissions, as well as the end of low-priced oil, photovoltaics will surely take center stage for the simple reason that there is no other nonpolluting energy option that works as effectively almost everywhere, from high above the Arctic Circle to Argentina, in South Africa as well as Scandinavia, for Russians and for Australians. Furthermore, a photovoltaic system does not have to intrude on uninhabited terrain or spawn miles of transmission lines, cluttering formerly open vistas as other generators of electricity must, since its placement can be confined to areas already in use by humans, such as rooftops.

In a deregulated electric market, photovoltaics has a much better chance to flourish than large solar mirror installations or wind machines. Steve Taylor, who worked on America's only power tower, explains: "With deregulation, you're going to get all sorts of power producers and suppliers having limited capital and a limited area in which to build power plants. When they want to add capacity to their grid, they're not going to have the money or the area, literally hundreds and hundreds of acres, that an effective power tower [or wind farm] requires, but they could find enough rooftops where photovoltaics could go up."[47]

Ever since the first solar cells traveled into space, photovoltaics has succeeded because the technology has always vied with the continually rising expense of delivering electricity to nonwired consumers. In contrast, other solar-generating technologies, such as the power tower and trough reflectors, have floundered because of their need for centralization, which has forced them to compete with large generators of power that benefit from the continually falling price of fossil fuels. The beauty of photovoltaics is that it requires no central plant or delivery lines, and it can be tailored to any power need from milliwatt to megawatt. Photovoltaics, for example, allows an anthropologist to run a laptop computer in the Orinoco wilderness with the same ease and cost effectiveness as the transit authority in Las Vegas can run security lighting for its bus shelters.

Photovoltaics is on the threshold of becoming a major energy source. Don Osborn of the Sacramento Municipal Utility Department notes, "Pho-

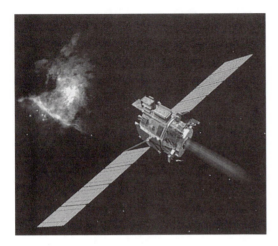

*Solar cells not only provide electricity for satellites, they now also provide the power that propels space vehicles.*

tovoltaics has finally reached the point where, within a reasonable time frame—in a decade or less—we can clearly expect solar cells to provide electricity on a widespread scale."[48] What the technology has accomplished so far presents but an inkling of things to come. The industry has barely tapped the vast markets for its products, like the 2.5 billion people who still live without electricity. Once those selling photovoltaics make serious inroads into this potential customer base, "We'll look back in time and say [that] a 10-megawatt plant, the average size of today's manufacturing facility, was an interesting pilot line," predicts Chris Schering, an international photovoltaics consultant.[49]

In 1956, publicists at Bell Laboratories made a bold prediction: "The ability of transistors to operate on very low power . . . gives solar [cells] great potential and it seems inevitable that the two Bell inventions will be closely linked in many important future developments that will profoundly influence the art of living."[50] Already the tandem use of transistors and solar cells in satellites, navigation aids, microwave repeaters, and televisions, radios, and cassette players in the developing world—and a myriad of other applications—has turned the Bell prediction to fact. It takes no wild leap of imagination to expect the transistor/solar cell revolution to continue until it encompasses every office and home in the world.

Thanks to solar cells powering devices from space to earth, people everywhere will enjoy the benefits of electricity without doing harm to their home, planet earth.

## Notes & Comments

1. Interview with Bernard McNelis.
2. Shell Solar, "Connecting You to the Sun." Brochure.
3. Interview with Steve Allison.
4. Interview with Terry Hart.
5. "Solar Energy is a Key to Development," *Nature* 281 (20 September 1979): 167.
6. Interview with Steve Allison.
7. "Solar Energy," *Shell Reports* (February 1978).
8. "Photovoltaics: A Question of Efficiency," *EPRI Journal* 6 (December 1981): 46.
9. T. Bhattacharya, "Solar Photovoltaics: An Indian Perspective, Solar Cells," *Solar Cells* 6 (1982): 258.
10. H. Khatib, "Electrification for Developing Countries," *EPRI Journal* 18 (September 1993): 29.
11. Interview with Steve Allison.
12. Interview with Terry Hart.
13. G. Foley, *Photovoltaic Applications in Rural Areas of the Developing World* (Washington, D.C.: The World Bank, 1996), xiii.
14. Interview with Jim Trotter.
15. United Nations, *Energy Issues and Options for Developing Countries* (New York: Taylor & Francis, 1989), 201.
16. Interview with Richard Hansen.
17. "Alleviating Poverty through Technology," *Science* 282 (16 October 1998): 410.
18. Interview with Bill Yerkes.
19. Ibid.
20. United Nations, *Energy Issues and Options*, 202.
21. The Italian government also has decided to light archaeological parks with photovoltaics. Its choice stems from the fact that this "constitutes a typical case where photovoltaics is cost effective because of the distance of the [parks] from the power grid. Moreover . . . the installation of distribution lines or conventional diesel generators is inadvisable, due to their unacceptable environmental impact." L. Barra et al., "Photovoltaic Systems for Archaeological Areas," *Tenth E. C. Photovoltaic Solar Energy Conference* (Lisbon, Portugal, 8–12 April 1991) (Dordrecht: Kluwer Academic Publishers, 1991), 805.
22. Interview with William Gould, Edison International.
23. Interview with Bill Mahoney.
24. Interview with John Thornton.
25. Interview with Jim Trotter.
26. Interview with Roland Skinner.
27. Ibid.

28. Interview with Ken Stokes.

29. L. Lamarre, "Renewables in a Competitive World," *EPRI Journal* 20 (November/December 1995): 17.

30. Interview with Peter Lawley.

31. Interview with Jim Tworpe, President, GPU Solar.

32. "The Best of What's New—100 of the Year's Greatest Achievements in Science and Technology," *Popular Science* 253 (December 1998): 77.

33. "Preface," *Annual Review of Energy* 14 (1989): v. The latest scientific evidence suggests that global warming that is due at least in part to human activity has affected the earth for over a century. T. Wigley et al., "Anthropogenic Influence on the Autocorrelation Structure of Hemispheric-Mean Temperatures," *Science* 282 (27 November 1998): 1676–79. Study after study published over the last decade have further convinced the majority of scientists, as the prestigious National Academy of Sciences warned President George Bush, "that global warming may well be the most pressing international issues of the next century." J. Kahn, "Global Warming and Energy Efficiency," *Sunworld* 14, no. 21 (1990): 44–45.

34. Interview with Mark Trexler.

35. For example, lighting homes with kerosene emits from three to six tons of $CO_2$ over twenty years. If a utility sponsored the widespread use of solar energy as the replacement fuel for kerosene, it would eliminate this $CO_2$ contribution and therefore qualify as a $CO_2$-offset project.

36. World Energy Council, *New Renewable Energy Resources* (London: World Energy Council, 1994), 131.

37. It must be noted that every other nonproducer of greenhouse gases presents major obstacles to its widespread future use. It is true that nuclear power does not produce greenhouse gases, but long before Three Mile Island and Chernobyl became synonyms for toxic disaster, scientists concerned about climate change ruled out nuclear because of "the serious environmental effects of its by-products." Study of Critical Environmental Problems (SCEP), "Climatic Effects of Man's Activities," *Man's Impact on the Global Environment* (Cambridge, MA: MIT Press, 1970), 12. The accidents at Three Mile Island and Chernobyl "intensified an already skeptical public attitude toward nuclear power and significantly reduced the potential of conventional nuclear power to contribute to the world's electricity in the coming decades." "Preface," *Annual Review of Energy* 15 (1990): v. The Chernobyl disaster ranks as the deadliest single human-caused catastrophe in the history of civilization, excluding, of course, acts of war. At the minimum, six thousand deaths can be directly attributed to Chernobyl. D. Marples, May/June, 1996, "The Decade of Despair," *The Bulletin of the Atomic Scientists* (May/June 1996): 25. More than a decade after the accident, the death toll from Chernobyl continues to mount

with "the explosive increase in childhood thyroid cancer in . . . the countries most contaminated by the 1985 Chernobyl nuclear accident." Michael Balter, "Chernobyl Thyroid Cancer Toll," *Science* 270 (15 December 1995): 1758. Most frightening for Europeans is that increased rates of thyroid cancer have been found as far as five hundred kilometers (three hundred ten miles) from Chernobyl "and in Europe everyone lives within five hundred kilometers of a nuclear power station." Balter, "Chernobyl Thyroid Cancer Toll." The peaks of the Alps still contain dangerous levels of radiation spewed by the disaster. "Radiation Still Present in Alps," *Santa Barbara News Press* (3 May 1998): A16. Nuclear power plants also make excellent targets for terrorists, as well as enemies at war. Increasing the number of nuclear plants makes the possibility of such disasters more imminent.

Wind energy and hydropower do not add greenhouse gases to the atmosphere either. Windpower presently generates electricity more cheaply than photovoltaics—where reliable winds blow. Large parts of the world, however, lack that resource. In addition, the proliferation of large windfarms would visually pollute open terrain and seascapes. P. Burgess and P. Pymn, *Solar Pumping in the Future; A Socioeconomic Assessment* (Pentyrch: CSP Economic Publishers, 1985), 23. Hydropower can only work where rivers and streams flow powerfully enough to move turbines. However, large-scale hydro projects can create significant social and environmental discord, including population displacement, riparian habitat destruction, and general havoc for all living downstream. Weather events are also a consideration. In Sri Lanka, for example, where hydro accounts for 93 percent of the island's electricity, the very dry weather of 1992 brought four months of power rationing. L. Gunaratne, "Solar Photovoltaics in Sri Lanka: A Short History," *Progress in Photovoltaics* 2, no. 4 (1994): 308. Changes in the world's hydrologic cycle, predicted by global-warming models, severely threaten the reliability of this power source in the future.

Power plants using mirrors—power towers and trough reflectors—that concentrate solar energy onto a boiler do not emit carbon dioxide, but they only work in cloudless weather. This rules them out for much of the world. They would work well in the Mojave desert or the deserts of Peru, but in central and northern Europe, "Forget it!" exclaimed Markus Real. Interview with Markus Real. Even in Spain and Sicily, poor winter weather can stymie power towers' and trough reflectors' ability to work. Wolfgang Palz, the project leader for the first power tower built in Sicily, learned this the hard way. Both the plant in Sicily and the one in southern Spain, optimal sites for Europe, "had lots of trouble with clouds. The weather gets really tough from December to April. Everywhere bad. This is the main reason solar thermal does not work in Europe. We just don't have the climate." Interview with Dr. Wolfgang

Palz. It should be kept in mind that large reflector plants situated in certain locations would cause irreparable harm to fragile ecosystems.

38. D. James Baker, "Special El Niño Weather Summary Issued," *United States Department of Commerce News*, National Oceanic and Atmospheric Administration, NOAA-98-019, 7 April 1998.

39. Personal correspondence, James G. Le Vere, Manager, Special Projects, Burlington Northern Santa Fe.

40. Interview with Jennifer Lee Bair-Riedhart of NASA. Solar-powered weather surveillance airplanes may also prove superior to satellites for transmitting information. Their lower altitude and capability to hover directly over communication centers on earth would allow for the transmission of more information faster and more cheaply than any satellite could deliver. Such aircraft would eliminate the need for an expensive rocket and fuel, as well as eliminating the use of ozone-destroying chemicals contained in rocket propellants. Additionally, if a problem occurred with the equipment, the aircraft could return to earth for repairs, as well as for upgrades. The aircraft would therefore not only be reusable, but it could always be equipped with the latest gear, unlike the current situation in which we are locked into antiquated technology aboard satellites which cannot return to earth.

41. "Hurricanes Georges and Mitch: PV systems fared well in the Dominican Republic and Honduras," ENERSOL NEWS (Winter 1999).

42. W. Young, "Preface," *Photovoltaic Applications for Disaster Relief,* FSEC-CR 849–95 (Cocoa, FL: Florida Solar Energy Center, 1995).

43. I. Melody, "Sunlight after the Storm," *Solar Today* (November 1992): 12.

44. C. Campbell and J. Laherrére, "The End of Cheap Oil," *Scientific American* 278 (March 1998): 78–79. Campbell and Laherrére base their predictions on the work of the late M. King Hubbard, who demonstrated that for any significant deposit of a finite resource, its unchecked removal follows the pattern of a bell curve, peaking when about half of the material is gone. Using the curve to tell the future of American oil production, Hubbard informed his employers at Shell Oil in 1956 that the amount of petroleum extracted in the United States would peak in 1970. Sure enough, 1970 came and American oil production began its decline. Other large oil fields have also followed King Hubbard's predictions. Although petroleum could be mined from shale deposits, sludge, and tar sands, not only would this cost much more, its extraction and use would pile new environmental problems onto the old. Exploiting shale and tar sands calls for strip mining. To extract the oil adds pollutants to the air, including greenhouse gases. Oil sludge contains heavy metals and sulfurs that would poison the atmosphere. Campbell and Laherrére, "The End of Cheap Oil." Switching to synthetic fuels would also hasten climate change as their manufacture and use release many of the gases implicated in global

warming. G. Macdonald, "Impact of Energy Strategies on Climate Change," *Preparing for Climate Change; A Cooperative Approach* (North American Conference on Preparing for Climate Change, Washington, D.C., 27–29 October 1987) (Rockville, MD: Government Institutes, 1988), 217.

45. Interview with Dr. Reinholdt Gregor, Shell International, Hamburg, Germany.

46. T. Beardsley, "Turning Green," Science and Business, *Scientific American* 271 (September 1994): 97.

47. Interview with Steve Taylor, Edison Technology Solutions.

48. Interview with Don Osborn.

49. Interview with Chris Schering.

50. Bell Systems press release, September 1956. AT&T Archives, Box #1960301, Warren, NJ.

# Index

# Illustration Credits

**Chapter 1.** *Page 2: top: The Wonderful Century: Its Success & Failures* by Alfred Russel Wallace (London: Swan Sonnenschain, 1898), page 2; **bottom:** Frontispiece from *Chaleur Solaire* by Augustine Mouchot (Paris: Gauthier–Villars, 1869). *Page 3: top:* Reprinted from *Memoirs of the American Academy in Rome,* Vol. XXIV, 1956, "The Open Rooms of the Terme del Ford at Ostia," by Edwin Daisley Thatcher; **bottom:** S. P. Langley, "Researches on Solar Heat" (Washington, D.C.: U.S. Government Printing Office, 1884), 166–68. *Page 4:* Butti–Perlin Archives, Santa Barbara, California. *Pages 5 & 6:* From John Ericsson, *Contributions to the Centennial Exposition* (New York: Printed by the author, 1876). *Page 7:* Butti–Perlin Archives. *Page 8: top:* Reprinted from *Engineering News*, 13 May 1909. ©McGraw-Hill Companies Inc. All rights reserved; **bottom:** Butti–Perlin Archives. *Page 10:* Pilkington Solar International GmbH, Köln, Germany. *Page 11:* Solar Power Corporation/Solar Power Limited. Courtesy Clive Capps. **Chapter 2.** *Page 16:* Courtesy of the Institute of Physics, United Kingdom. *Page 17:* Reprinted by permission of the CINDEX. Add. 7657, R51063. Cambridge University Library, Archive and Manuscript Collection, University College London. *Page 18: top: From Scientific and Technical Papers of Werner von Siemens, 1892–1895*, vol. 1 (London: J. Murray, 1892); **bottom:** From Thomas Benson, *Selenium Cells* (New York: Spon and Chamberlain, 1919). *Page 19:* Letter to Sir Oliver Lodge, 28 May, 1887, MS Add 89/172, University College London Library. **Chapter 3.** *Pages 26, 27, 31, 32 (top two photos):* Property of AT&T Archives, Warren, New Jersey. Reprinted with permission of AT&T. *Page 29:* United States State Department; *Page 32: bottom:* Edison Technology Solutions. **Chapter 4.** *Page 36 & 37 (top photo):* Property of AT&T Archives, Warren, New Jersey. Reprinted with permission of AT&T. *Page 37: bottom:* From the collection of Donald E. Osborn. Original photograph by William A. Rhodes. *Page 39: top:* Courtesy Dr. Martin Wolf; **bottom:** Courtesy Peter Iles. Hoffman Electronics Corporation in-house publication. **Chapter 5:** *Page 42:* Courtesy Frederica Meindl. *Page 43:* Reprinted with permission from *Popular Science Magazine.* ©1956 Times-Mirror Magazines Inc. *Page 45:* 11th Annual Battery Research and Development Conference Proceedings, Power Sources Division, United States Army Signal Engineering Laboratories, Ft. Monmouth, New Jersey, 1957. *Page 46:* Courtesy Peter Iles. Hoffman Electronics Corporation in-house publication. **Chapter 6.** *Pages 50 & 51:* Courtesy Peter Iles. Hoffman Electronics Corporation in-house publication. *Page 55:* Courtesy Elliot Berman. Solar Power Corporation/Solar Power Limited. **Chapter 7.** *Page 58.* Courtesy Elliot Berman. Solar Power Corporation/Solar Power Limited. *Page 59:* Courtesy Automatic Power, Inc. *Page 61:* Graphic furnished by Tideland Signal Corporation. *Page 63: left:* Courtesy Siemens Solar Industries; **right:** Solar Power Corporation/Solar Power Limited. Courtesy Clive Capps. *Page 64:* Good-All Manufacturing, Inc., a Corrpro Company. *Page 65:* Solar Power Corporation/Solar Power Limited. Courtesy Clive Capps. **Chapter 8.** *Pages 73 & 74:* Courtesy Lloyd R. Lomer, Captain, USCG Ret. *Page 75: top:* Photograph provided by Solarex; **bottom:** Courtesy Automatic Power, Inc. **Chapter 9.** *Page 78:* Solar Power Corporation Solar/Power Limited. Courtesy Clive Capps. *Pages 79 & 81:* Courtesy Norfolk Southern Corporation. **Chapter 10.** *Page 88:* Courtesy Peter Iles. Hoffman Electronics Corporation in-house publication. *Pages 91 & 93:* Courtesy John Oades. *Page 92:* Courtesy GTE Corporation. *Page 94:* Reprinted with permission from April 16, 1979, issue of *Telephony.* Copyright 1979, Primedia Intertech, Overland Park, Kansas. All rights reserved. *Pages 97, 99, 100 & 101:* Provided by the Telstra Corporation Limited (known in Australia as Telstra). **Chapter 11.** *Pages 107, 112, 113, 114 & 122:* Mali Aqua Viva,

Pere Verspieren au Mali: "Periya Bugu, B.P 1, San, Mali" et "Sahel Aqua Viva," c/o Fondation de France, 40 Ave Hoche 75008, Paris, France. *Page 108:* Solar Power Corporation/Solar Power Limited. Courtesy Clive Capps. *Page 109:* Courtesy Dominique Campana, director of International Affairs, ADEME, Paris, France. *Page 110:* Courtesy Bernard McNelis, managing director, IT Power Ltd, The Warren, Bramshill Road, Eversley, Hampshire, RG270PR, United Kingdom. *Page 120:* Courtesy Siemens Solar Industries. **Chapter 12.** *Page 131: left: Energie Pacifique, l'energie, l'habitat et le soleil. A Tahiti et sous les Tropiques.* Programme Territorie des le Polynésie française—CEA—AFME. Courtesy Clive Capps; ***right:*** *Les Energies Renouvelables en Polynésie Française (Renewable Energy in French Polynesia).* Programme Territorie des le Polynésie française—CEA—AFME. Courtesy Patrick Jourde. *Page 133:* Courtesy Lalith Gunaratne. *Page 135:* Courtesy Mark Hankins, solar energy consultant based in Nairobi, Kenya, working for Energy Alternative AFRICA Ltd. and Energy for Sustainable Development Ltd. His book *Solar Electric Systems for Africa* is available by contacting Energy Alternatives AFRICA, PO Box 76406, Nairobi, Kenya. *Page 136: top:* Courtesy Richard H. Acker, 1994; ***bottom:*** Courtesy Mark Hankins. *Page 137: top:* Courtesy Mark Hankins; ***bottom:*** Courtesy Richard H. Acker. *Page 138:* Courtesy Richard Acker. *Page 140:* Courtesy Soluz Inc. *Pages 143 & 144:* Courtesy Solar Electric Light Company (SELCO). **Chapter 13.** *Page 150: top:* National Science Foundation; ***bottom:*** Reprinted with permission from Wolfgang Palz, *Solar Electricity: An Economic Approach to Solar Energy* (London: Butterworth, 1978). ©UNESCO. *Page 151:* Courtesy Alpha Real. *Page 153:* Courtesy Siemens Solar Industries. *Page 155:* Photo by Donald E. Osborn, Sacramento Municipal Utility District. *Pages 157, 160 & 161:* Pilkington Solar International Gmbh, Köln, Germany. *Page 162:* Courtesy Kiss + Cathcart, Architects; Sun Hung Rendering; Fox & Fowle, Basebuilding Architects; The Durst Organization, Developer. **Chapter 14.** *Page 166: top left:* Courtesy Peter Iles. Hoffman Electronics Corporation in-house publication; ***middle right:*** Courtesy Siemens Solar Industries; ***middle left:*** Courtesy United Solar System Corporation; ***bottom:*** Courtesy John P. Thornton, National Renewable Energy Laboratory. *Page 167:* Courtesy Donald E. Osborn, Sacramento Municipal Utility District. *Pages 168 & 170:* Courtesy Mobil Oil Corporation. *Page 169:* Photograph supplied by Solarex. *Page 172:* Courtesy ASE Americas Inc. *Page 174:* Courtesy ENTECH. *Page 175:* From the collection of Donald E. Osborn. Original photograph by William A. Rhodes. *Page 177:* Courtesy Dr. David Carlson, Solarex, a business unit of AMOCO/Enron Solar. *Page 179:* Courtesy United Solar Systems Corporation. *Page 181:* Courtesy AstroPower. A. M. Barnett, R. B. Hall, and J. A. Rand, "Thin Polycrystalline Solar Cells," *NRS Bulletin* (October 1993): 33–36. *Pages 183 & 184:* Reprinted with permission of Martin Andrew Green. Not to be reproduced without permission of copyright holder. **Chapter 15:** *Page 191: top:* Courtesy B. D. Agchbayar, Ulan Bator, Mongolia; ***bottom:*** Photograph provided by Solarex. *Page 192:* Provided by the Telstra Corporation Limited (known in Australia as Telstra). *Page 193:* Reprinted with permission of Iridium LLC. *Page 194:* Courtesy Siemens Solar Industries. *Page 195:* Courtesy Alpha Real. *Page 196:* Courtesy Roland Skinner, Northwest Rural Public Power District (Nebraska). *Page 197:* Photograph provided by Solarex. *Page 198:* Prepackaged MAPPS system installed by Solar Electric Specialties, a division of Applied Power Corporation. *Page 199:* Courtesy United Solar Systems Corporation. *Page 200:* Courtesy National Aeronautics and Space Administration, Dryden Flight Research Center, PO Box 273, Edwards, California 93523. *Page 201:* Courtesy United Solar Systems Corporation. *Page 202:* Courtesy Shell Solar Energy bv. *Page 204:* Courtesy NASA/JPL/Caltech.